KU-657-109

Statistics and Analytical Methods in Automotive Engineering

Conference Organizing Committee

S Edwards
Ricardo Consulting Engineers Limited, UK

D Grove
Independent Consultant, UK

D Fish
Jomega, UK

J Read
Independent Consultant, UK

M Gidlow
Retired – Formerly Ford Motor Company, UK

D Rose
Ford Motor Company, UK

IMechE
Conference Transactions

I MECH E

International Conference on

Statistics and Analytical Methods in Automotive Engineering

24–25 September 2002
IMechE HQ, London, UK

Organized by
The Combustion Engines and Fuels Group in conjunction with the Automobile Division of the Institution of Mechanical Engineers (IMechE)

Sponsored by
Ricardo Consulting Engineers Limited

Co-sponsored by
EIS
Royal Statistical Society

IMechE Conference Transactions 2002–4

**Professional
Engineering
Publishing**

Published by Professional Engineering Publishing Limited for The Institution of Mechanical Engineers, Bury St Edmunds and London, UK.

First Published 2002

This publication is copyright under the Berne Convention and the International Copyright Convention. All rights reserved. Apart from any fair dealing for the purpose of private study, research, criticism or review, as permitted under the Copyright, Designs and Patents Act, 1988, no part may be reproduced, stored in a retrieval system, or transmitted in any form or by any means, electronic, electrical, chemical, mechanical, photocopying, recording or otherwise, without the prior permission of the copyright owners. *Unlicensed multiple copying of the contents of this publication is illegal.* Inquiries should be addressed to: The Publishing Editor, Professional Engineering Publishing Limited, Northgate Avenue, Bury St Edmunds, Suffolk, IP32 6BW, UK. Fax: +44 (0) 1284 705271.

© 2002 The Institution of Mechanical Engineers, unless otherwise stated.

ISSN 1356–1448
ISBN 1 86058 387 3

A CIP catalogue record for this book is available from the British Library.

Printed by The Cromwell Press, Trowbridge, Wiltshire, UK

UNIVERSITY OF HERTFORDSHIRE
HATFIELD CAMPUS LRC
HATFIELD AL10 9AD

BIB
438775
CLASS 629 252 INT
LOCATION MAIN STANDARD
BARCODE 600062008X

The Publishers are not responsible for any statement made in this publication. Data, discussion, and conclusions developed by authors are for information only and are not intended for use without independent substantiating investigation on the part of potential users. Opinions expressed are those of the Author and are not necessarily those of the Institution of Mechanical Engineers or its Publishers.

Related Titles of Interest

Title	Editor/Author	ISBN
Statistics for Engine Optimization	S P Edwards, D M Grove, and H P Wynn	1 86058 201 X
Vehicle Handling Dynamics	J R Ellis	0 85298 885 0
Advances in Vehicle Design	J Fenton	1 86058 181 1
Handbook of Automotive Powertrain and Chassis Design	J Fenton	1 86058 075 0
IMechE Engineers' Data Book – Second Edition	C Matthews	1 86058 248 6
Integrated Powertrains and their Control	N D Vaughan	1 86058 334 2
Total Vehicle Technology 2001	P R N Childs and R K Stobart	1 86058 324 5
Vehicle Systems Integration	A V Smith and C Hickman	1 86058 262 1
Vehicle Systems Integration in the Wired World	A V Smith and B J Wilde	1 86058 343 1
Integrated Powertrain Systems for a Better Environment	IMechE Conference	1 86058 224 9
Vehicle Noise and Vibration	IMechE Conference	1 86058 386 5

For the full range of titles published by Professional Engineering Publishing contact:

Marketing Department
Professional Engineering Publishing Limited
Northgate Avenue
Bury St Edmunds
Suffolk
IP32 6BW
UK

Tel: +44 (0)1284 724384
Fax: +44 (0)1284 718692
E-mail: orders@pepublishing.com
Website: www.pepublishing.com

Contents

Vehicles, Systems, and Components

Statistics and Analytical Methods in Automotive Engineering

INTRODUCTION

Interest in the application of statistical and analytical methods in automotive engineering has increased dramatically since 1998, when the first Institution of Mechanical Engineers event relating to 'Statistics for Engine Optimization' took place. Awareness of the benefits from the application of statistical experimental design, modelling, and optimization is now widespread and the use of analytical methods (in the form of computer aided engineering) essential. Nevertheless, we felt it appropriate to revisit these complementary techniques, firstly with a broader vision, considering not just engine optimization but the whole vehicle development and, secondly, from the product delivery process perspective, where significant gains can be made by utilizing information in an efficient manner. Consequently, this second event has been organized: a conference on 'Statistics and Analytical Methods in Automotive Engineering', the proceedings of which are given here.

OVERVIEW

On the following pages you will learn about new developments in statistical and analytical techniques and their application from engine research right through to vehicle warranty analysis.

Regarding *powertrain development* there is a clear message relating to the benefits of the application of the techniques to improve powertrain design and calibration robustness (see for example Bates and Heikal and Shayler *et al.*). Considering calibration in particular, for emissions, performance and/or driveability optimization, we find many new applications of the techniques (for example Roudenko *et al.*, Roepke *et al.* and Pickering *et al.*). These applications are being supported by the development of new tools for the test bed and desktop (see for example Morton and Knott and Raynaud *et al.*) and new techniques such as two stage regression and stochastic process methods (as reported by Rose *et al.* and Seabrook *et al.*). There is even consideration about how to use existing information in an efficient manner (always an issue when starting a new development cycle) using Bayesian statistics (Ward *et al.*).

Regarding *vehicle systems and components* there are themes related to vehicle structures and dynamics, component manufacture through to in-service reliability. The analysis and optimization of vehicle structural design is considered using dimensional variation (Koganti and Zaluzec) and reliability (Yang *et al.*), with particular attention given to vehicle crash analysis (Le Glatin *et al.*) and dynamics (Fothergill and Allman-Ward). Production variability and reliability is analysed (see for example Linsley *et al.* or Athavale *et al.*) and, via six sigma techniques, studied in relation to design (Brunson *et al.*) through to complete vehicle development. Once in the field, component reliability and its possible effects are considered (by Campean and Brunson and Krivtsov *et al.*) and how vehicle performance might be improved via predictive algorithms (by Borgarello *et al.*).

CONCLUSION

We hope you find this second collection informative and stimulating and, as the increased breadth of the applications attests, encouraging enough to persuade you to try the techniques for yourself. Once more, we would welcome your comments and questions and look forward to reporting further developments on the subject in the future.

Dr Simon Edwards
Mr David Fish
Mr Mike Gidlow
Dr Dan Grove
Eur Ing Jack Read
Dr Dean Rose

Engines

C606/020/2002

Bayesian statistics in engine mapping

M C WARD, C J BRACE, N D VAUGHAN, and **G SHADDICK**
University of Bath, UK
R CEEN
CP Engineering, UK
T HALE and **G KENNEDY**
Cosworth Technology, UK

ABSTRACT

In order to comply with increasingly stringent emissions standards and meet driveability requirements, modern automobile engines are equipped with an increasing number of subsystems and controlling elements. The result has been to greatly increase the calibration effort required to find the parameter settings that offer the best global compromise across the entire engine map. An increasing trend within engine testing is the application of statistical tools to minimize the test time required for calibration; these include design of experiments (DoE), Bayesian and stochastic methods, which have all proved themselves in addressing multivariate problems in other fields.

This paper examines a core subset of engine mapping termed 'spark sweeps'. Spark sweeps are necessary to identify the operating regions that are potentially destructive to the engine, and to provide torque reduction data necessary for satisfactory transient operation. Software for automating spark sweeps is currently running on many test rigs. One such example is used as the basis for this work.

This paper shows a further improvement in terms of time by coupling a Bayesian prior knowledge algorithm to the spark sweep software. The benefits found are likely to be applicable to the wider calibration problem. This development was carried out on a turbocharged gasoline engine rig controlled by a commercial host system, the direction of which was abstracted into the MATLAB/SIMULINK environment. This link was developed in house at the University of Bath and enables the full range of MATLAB data processing, statistical, and optimisation routines to be used online. This paper presents the Bayesian prior knowledge algorithm itself, and shows a potential improvement in efficiency of the whole spark sweep procedure.

ABBREVIATIONS

BLD   Borderline Detonation
EGT   Exhaust Gas Temperature
EMS   Engine Management Systems
ETC   Electronic Throttle Control
ETC   Electronic throttle control
GDI   Gasoline Direct Injection
ICAM   Integration Calibration and Mapping Component
TMAX   Maximum Torque
ME7   Bosch ME7 Engine Management system
RL   Relative Load – "Relativ Luft"

1. INTRODUCTION

In order to comply with increasingly stringent emissions standards and to meet driveability requirements, modern engines are equipped with an increasing number of subsystems and controlling elements. This has led to the use of torque based engine control architectures [1], which use electronic throttle control (ETC). The result has been to greatly increase the calibration effort required to find the parameter settings that offer the best global compromise across the entire engine map. This task is set to get significantly harder in the future with the increasing adoption of gasoline direct injection (GDI) engines and the advances in diesel technology.

A current trend within engine testing aimed at reducing the time required for the calibration procedure, is the application of statistical tools developed for use in other fields where empirical models of systems are necessary. This paper examines the potential Bayesian Statistics [2] which offers a way to reuse prior knowledge that is traditionally discarded by classical statistical methods such as design of experiments [3]. Data yielded from experimentation is used to continually update a model, which initially is dominated by prior knowledge, then progressively becomes influenced by experimental results. This means that a test program can be guided in real time by the feedback of data, and hence should be more flexible than the traditional approach of collecting all the data before building the model. This paper seeks to examine the effectiveness of a Bayesian Statistics algorithm at characterizing a particular area of engine behaviour by comparison against a traditional method of collecting the same data. Although the experimentation outlined in this paper is centered on spark mapping it is envisaged that the same algorithm could be developed to tackle the wider engine calibration problem.

1.1. Spark sweeps

The power output of a spark ignition engine is determined by the airflow through it. This is governed via the throttle valve; and in the case of boosted engines also a turbocharger. The airflow to the inlet manifold is normally determined using a hot film sensor located upstream of the throttle assembly and is processed by the ECU into a measure designated 'Relativ Luft' meaning literally 'relative air' but more commonly translated as Relative Load (RL). RL is defined from 0 to 130% with values over 100% meaning that the engine is running at wide-open throttle and that the airflow is being boosted by the turbocharger. Knowledge of the relationship between ignition timing and torque at a fixed RL is necessary for calibration of

C606/020/2002 © With Authors 2002

gasoline engines with a torque based electronic control unit (ECU) because ignition timing is used as a fast control path for traction control, during rapid pedal transients and for knock control. Typically during a spark sweep engine speed, lambda and RL are controlled to constant values with the only variable being ignition timing.

1.2. Testbed Layout and the spark sweep software
The test cell used for this work is based upon a Volkswagen group production specification 1.8 turbocharged gasoline engine. This unit has a maximum power output of 135kW and a maximum torque of 210 Nm. The engine is managed with a torque based ECU. The engine is loaded by a Schenck eddy current dynamometer, capable of absorbing up to 230 kW. Control of the test cell, dynamometer and data acquisition is carried out by the CP Engineering Cadet V12 software [4].

The test bed software network is shown in Figure 1. It comprises of a testbed control PC which acts as server to 2 client PCs which run the calibration tool [5] (ETAS-INCA) and MATLAB respectively. The connections are implemented by an in house TCP/IP Ethernet link [6]. This arrangement means the testbed control PC handles time critical tasks (testbed control, limit monitoring, and parameter settling) and gives MATLAB authority over ECU parameters under supervision of the testbed control PC.

ICAM (Integrated Calibration and Mapping) is a software component [7] Figure 2. It runs in Cadet to automatically acquire spark sweep data for torque based engine management systems. It allows torque data to be measured for any required combination of engine control parameters. The program consists of two nested procedures.

1. Inner loop - spark sweep, this consists of a forward spark sweep where the ignition angle is advanced stepwise and the torque logged until maximum torque (TMAX) or border line detonation (BLD) is found, then a reverse sweep is carried out with the ignition being retarded until either misfire or EGT limits are exceeded. In both cases ICAM terminates the sweep immediately to prevent hardware damage. For this work the knock indication is provided from accelerometer signal processed by the ECU and fed into the Cadet via ASAP3. Misfire was indicated by a signal representing an increased level of fluctuation in crankshaft speed and EGT via thermocouples in the exhaust manifold. This effect of the ignition angle on the engine torque is plotted for the user on the ICAM screen as shown in Figure 2, and the relationship may be represented by a second order curve. The horizontal line indicates TMAX.

2. Outer loop - predefined engine parameter settings, at which a spark sweep is to be carried out. For example engine speed, relative load, AFR, camshaft phase.

ICAM was run in its standard form to produce baseline surfaces of the engine for use as prior knowledge data sets. These surfaces were achieved by logging torque when with varying lambda and spark angle. 11 point Spark sweeps were carried at lambda values of 0.8, 1, and 1.2 at the following sites:

- Engine Speed (rev/min): 2000

- Relative Load (%): 20,30,50

Giving a total of 22 testpoints. The aim of the experimentation therefore is to obtain the same surface with a reduced number of test points. The form of the torque response and associated variance at 2000 rev/min and a relative load of 30% is shown in Figure 4.

2. BAYESIAN STATISTICS

The Classical or frequentist approach to statistics is to assume that nothing is known about a system until all the experiments on it are complete and a model is constructed. This method normally uses a random approach to the collection of data, thus ensuring the removal of systematic errors during testing. The limitations of this approach are that there is no feedback during the test program and data gathered from previous experimentation is discarded. Bayesian Statistics [2] on the other hand allows the use of prior information, which in theory should lead to a more efficient program of experimentation as all available data about the system are used and feedback can be provided to guide the test program whilst it is underway. This approach has been previously applied offline to engine test data in [8]. A Bayesian statistics approach offers the potential of outlier detection, owing to the existence of prior knowledge about the testing. Outlier detection can be used to flag up possible measuring device faults or other possible causes of spurious data.

The Bayesian algorithm begins with the formulation of a prior model derived via old test data and/or expert knowledge entered by the operator. In practice it would be difficult for the operator to generate a prior knowledge model without comprehensive prior test data. The operator will most likely only be able to 'tune' a prior model. In this work it is assumed that the engine data is normally distributed, hence at a given test point the prior knowledge provides a response distribution in the form:

$$Z_{prior} \sim (\mu, \sigma^2)$$

With mean (μ) and variance (σ^2). The engine response is then sampled at this test point giving an observed distribution:

$$Z_{obs} \sim (\mu, \sigma^2)$$

The Bayesian algorithm then produces a weighted average of these two results called the posterior:

$$Z_{post} \sim (\mu, \sigma^2)$$

thus prior information is combined with the observed test data. The posterior model then becomes the prior for the next test point. As previously stated this work assumes that the prior and observed data are normally distributed, which dictates that the posterior is also normally distributed. In the case where the prior and observed data are thought to be distributed differently the Bayesian updating algorithm becomes significantly more complicated. Examples of how to tackle such a scenario are given in [2].

Knowledge of the posterior variance will allow confidence intervals for the posterior model to be generated. These lead to 'predictive' intervals for the next test point such that outlier detection and test point placement can be optimised.

C606/020/2002 © With Authors 2002

2.1. Algorithm implementation

A Bayesian updating algorithm was developed in MATLAB [9] so it could be interfaced with the Test Bed Host system and allowed use of the modeling toolboxes provided. The structure of this algorithm is outlined schematically in Figure 3, which shows how the MATLAB code interfaces with the testbed and ECU. The method employed in this implementation uses separate models to hold the response and variance values over the range of operating parameters. A 'point' posterior mean and variance are computed when a new test point is sampled, and then the models are refitted with the posterior values included. This code was developed from a previous implementation described in [10] in which the actual model coefficients of a second order least squares model were updated. The performance of this method was found to be unsatisfactory due to difficulties assigning realistic variances to the coefficients.

The prior variance at a test point is calculated from the value of the prior and the sample response, as the further away the observed point is from the prior then it follows that the less confidence there is in the prior (assuming that this observation has been passed as plausible). It was also necessary to include in the prior variance the variance returned by the least square fitting subroutine. The observed variance was made equal to the variance given by the current prior variance model from the previous step as it is the only value of variance that can be assumed.

A weighted fit was needed to bias the models towards the posterior data, as the models were being fitted from a larger number of test points each time a new posterior was added. A least squares fit of the responses and one of the response variance is then applied to these matrices. It was found that the torque, spark lambda surfaces could be satisfactory fitted with a second order least squares fit, however for different data other models can just as easily be used. The implementation assumed that the variance of the coefficient terms was independent and normally distributed, as well as the prior and sample data from the test bed. The advantage of using a model where the likelihood, prior and posterior are all normal is that the normal distribution is time invariant.

The Bayesian algorithm communicates with the host system via a Ethernet link to select new test points and to subsequently measure the torque response of the engine online. The Bayesian technique is generic across different modeling techniques. Hardware protection of the engine is looked after solely in Cadet, if combustion knock, misfire of EGT limits were encountered as a result of the parameter settings the host system immediately imposed safe settings and returned a failure message to MATLAB allowing another point to be chosen. The MATLAB implementation includes an interactive display for the rapid evaluation of the algorithm potential. This allows the user to manipulate the prior knowledge in a drag and drop fashion and then to select the next test point manually by clicking on the plots.

Configurable parts of this algorithm, in addition to the type of model used, which will have an impact on the performance of the algorithm, are -

- The test point selection and model convergence criteria

- Prior Knowledge Quality

- Variance estimation

2.2. Convergence criteria

When rate of change of the model falls below a threshold as more posterior points are added, the algorithm is terminated as no more knowledge is being added to the model. As the model converges it should be seen that the computed posterior variance will tend towards the experimental accuracy characteristic of the testing itself. The strategies for determining when this occurs are numerous, and depend on the modeling technique being used and the accuracy requirement specific to the testing being carried out.

A threshold value is set for the variance to determine when the testing is terminated. This could be a blanket threshold or be biased towards critical areas for example the stoichiometric operating region of the engine. This threshold could also be the rate of change of the variance. In this work a blanket threshold for the variance of 1 was used for termination. This was kept constant to allow testing of other parts of the algorithm. Another possible convergence criterion could be rate of change of the posterior model coefficients, or possibly the most dominant model coefficients.

3. RESULTS

3.1. Prior knowledge quality

The quality of the prior knowledge impacts the number of test points required for convergence. Figure 5 and Figure 6 show testing of the algorithm at a target operating point of N=2000 rpm and RL=30%, where a known target model exists. Two data sets collected at N=2000 rev/min, RL=50% and at N=2000 rev/min, RL = 20% were used as examples of bad prior knowledge and good prior knowledge respectively. In the first quadrant of these figures there is a plot of the current posterior model (lower surface) against the known target model (upper surface).

It can be seen that the case using good prior knowledge has converged after only 7 test points, while the case using bad prior knowledge converges after 21 points, this is an improvement of the 33 testpoints where no prior knowledge is assumed. Using the Bayesian technique with inaccurate prior knowledge will not lead to an incorrect posterior model, as if enough test points are sampled an accurate model will still be created. This will however, be at the expense of an increased number of test points because the prior will incorrectly influence the model. For this reason significantly incorrect prior knowledge leads to an increased number of test points over the case where no prior knowledge is assumed.

3.2. Variance estimation

In order to obtain posterior distribution at a particular test point a measure of the prior as well as the observed point variance is needed. The variance describes the confidence that a given test point response is correct. As in most cases the prior knowledge is going to be from previous engine test data it is likely that a knowledge of the prior variance will be available from repeat points. Figure 4 shows a baseline map of the variance at a relative load of 30%. The variance in this case is dictated by the running stability of the engine, which degrades when running lean and retarded. The variance of the observed response was derived from is deviation from the expected response given by the prior information at that point.

Accurate modeling of variance for the prior model is important for quick and accurate convergence of the posterior solution. In the case where the variance is set too 'tight', the

C606/020/2002 © With Authors 2002

posterior is affected less by the observed data and more by the prior, meaning that more test points are needed until convergence. As the variance is loosened the observed data will have more and more influence leading to faster convergence, however the posterior model will be less tolerant of spurious data (outliers). Figure 7 and Figure 8 show the effect of an artificially generated spurious point on the posterior models for a 'tight' and a 'loose' variance case using good prior knowledge. In the first quadrant of these figures there is a plot of the current posterior model (lower surface) against the known target model (upper surface). In the case of the 'loose' variance the model is unduly influenced by the incorrect data and is thrown off track. This results in an increased number of test points being required to compensate for the incorrect inclusion of the spurious point. The algorithm is influenced less by the spurious point less in the 'tight' variance case, and will converge more quickly.

3.3. Testpoint selection strategies
Traditional engine calibration follows a fixed data selection procedure where test points are selected incrementally from a range of data, this is often done to prevent hardware damage for example in the case of combustion knock, or temperature limits. The test bed engineer would argue that in a carefully controlled test cell environment with proper parameter settling periods and temperature controls the need for randomization is minimized. As Bayesian statistics offers us the chance to continuously update our model it is possible to select test points one at a time and analyze the effect they have on the model. In this way information from each test point can be maximized and experimentation can be guided in order to increase model building efficiency. A number of test point selection strategies were investigated, however the results can only be qualitatively outlined here due to lack of space.

Random test point selection fulfills the statistical requirement of randomization, which underpins the assumption that the experimental errors are independent of each other. This method requires the most points for convergence as it leads of areas where duplicate data points are clustered which adds superfluous data to the model and areas where no data is added to the model.

Sequential test point selection makes no use of the posterior model as the data is sampled. A predefined test matrix is followed. This offers a quicker convergence of the posterior model than with random test point selection, but when good prior knowledge is used superfluous data is added to the model. In the case where the prior knowledge is uncertain and a large variance appropriate then this method would be ideal to ensure the design space is covered.

Maximum variance selection samples the next test point where the posterior model is uncertain, that is the region of highest variance. This method offers the fastest convergence of the posterior as the maximum amount of knowledge is added each update, and use is made of the model as soon as the data is returned. This method could follow on from the execution of a sparse grid pattern. This method will flag up earlier whether the prior knowledge is accurate, and would be useful for use with higher order models to ensure no local peaks of troughs in the response surface are missed.

4. CONCLUSIONS AND OUTLOOK

An algorithm based on the Bayesian updating method was developed to add prior knowledge into a standard calibration tool. The algorithm interfaced via Ethernet with this tool running on the host system, which performed limit monitoring and communication with the ECU.

It was found that the Bayesian algorithm if used with appropriate prior knowledge has the potential to reduce the number of test points required to build the model outlined in this work. This algorithm has been tested with data sets able to be modeled with a second order least squares fit. As more and more test points were added to the posterior model it was seen that the posterior variance tends towards the experimental accuracy characteristic of the testing itself. The use of accurate prior knowledge with a sensible estimation of variance is required for good convergence times. Using bad prior knowledge will result in more test points than no prior knowledge at all, because the model will be incorrectly influenced by the prior. As long as the modeling technique is capable of fitting the data the use of bad prior knowledge will not result in an incorrect model because as more test points are sampled the algorithm will become biased towards the sampled data.

The convergence criteria on which testing is terminated is the easiest part of the algorithm to configure, and depends on the required accuracy of the testing. Maximum variance selection is the most efficient test point selection method as this ensures that the maximum amount of information is derived from each new data point. The least obvious part is the assignment of the variance for the prior knowledge to the observed data. The definition of the prior variance influences the speed of convergence of the posterior model. If the variance bounds are set tight the posterior will be less affected by the sample data added to it and more by the prior. This will mean that if the prior knowledge is not very accurate then a large convergence time will occur. If on the other hand the variance of the prior is set very loose then the outlier detection of this method is lost.

This algorithm has been tested with data sets able to be modeled with a second order least squares fit, the testing is to be extended to higher orders. The Bayesian method is not specific to any particular model fitting technique. The next stage of work is to extend the Bayesian algorithm to modeling techniques other than least square fitting, for example natural neighbours or splines. This will allow more complicated calibration tasks to be tackled. The Bayesian Method also offers the potential for outlier detection, due to the existence of prior knowledge, and further work will include simulating measurement errors to access this potential.

REFERENCES

(1)   Bosch, 1999, "ME-Motronic Engine Management", 1st Edition Bosch.
(2)   Lee, 1994, "Bayesian Statistics, an Introduction", Edward Arnold.
(3)   Stuhler et al, 2002 "Automated Model Based GDI Calibration Adaptive Online DOE Approach", SAE Paper 2002-01-0708.
(4)   CP Engineering, 2000, "Cadet V12 – Run test manual".
(5)   ETAS GmbH, 1999, "INCA MATLAB Interface V1.2", Document AM010101 R1.2.1 EN.

(6)   Tiley, Ward, 2002, "MATCAD Ethernet Link Users Guide", University of Bath,
        www.bath.ac.uk/~enpmcw.
(7)   CP Engineering, 2000, "ICAM component manual".
(8)   Mowell, Robinson, Pilley 1996, "Bayesian Experimental Design and its Application to
        Engine Research and Development", SAE Paper 961157.
(9)   Mathworks, "MATLAB online help", www.mathworks.co.uk.
(10)   Finch, 2001 , "An Application of Bayesian Statistics to Engine Modeling.", MSc
        Thesis, Bath University.
(11)   Weltner, Grosjean, Schuster, Weber, 1994, "Mathematics for Engineers and Scientists",
        Stanley Thornes.

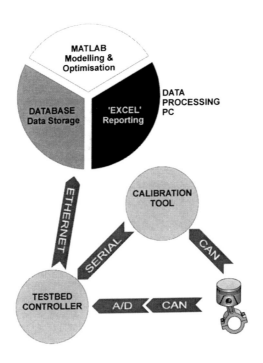

Figure 1: Testbed Network

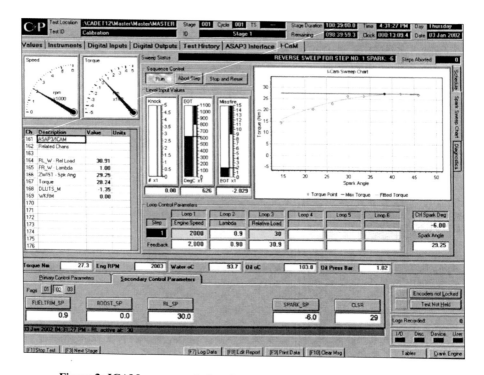

Figure 2: ICAM component, showing a Torque Spark data points with a second order fit

C606/020/2002 © With Authors 2002

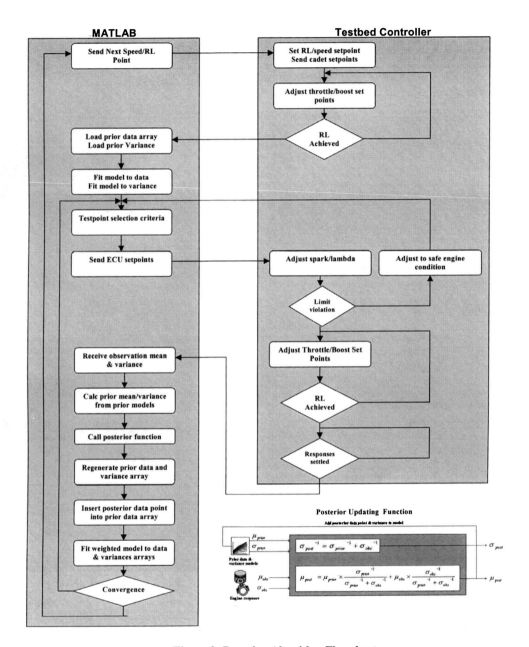

Figure 3: Bayesian Algorithm Flowchart

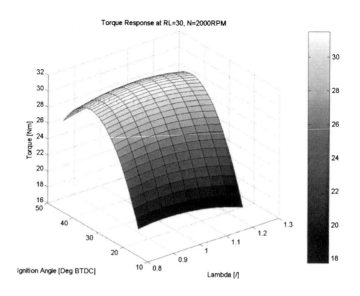

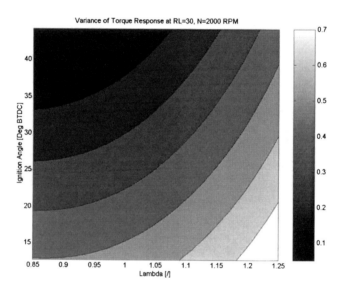

Figure 4: Base Line Torque Response and Associated Variance

C606/020/2002 © With Authors 2002

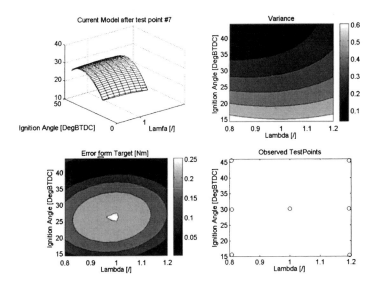

Figure 5: Use of good prior knowledge

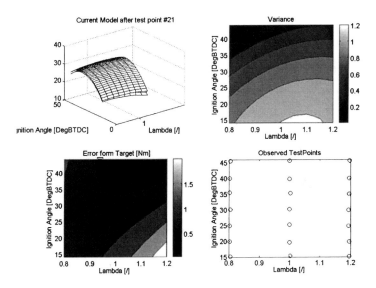

Figure 6: Use of bad prior knowledge

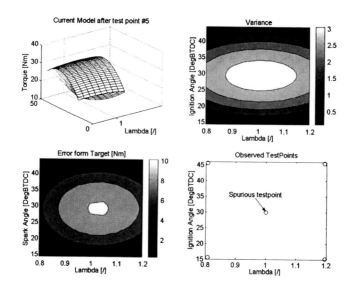

Figure 7: Effect of a spurious response on a model with a tight prior

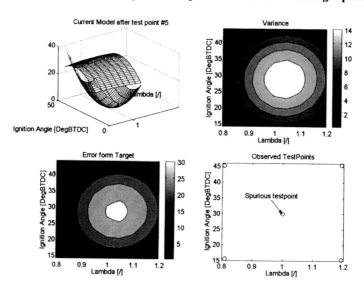

Figure 8: Effect of a spurious response on a model with a loose prior

C606/020/2002 © With Authors 2002

C606/028/2002

Applications of advanced modelling methods in engine development

J SEABROOK, B ROGERS, G FARROW, and J PATTERSON
Ricardo Consulting Engineers Limited, Shoreham-by-Sea, UK
S EDWARDS
Ricardo GmbH, Stuttgart, Germany

ABSTRACT

Design of Experiments (DoE) is now widely applied in engine development. However, there are some important engineering activities where it is difficult to apply classical DoE techniques comprising 2- & 3-level fractional designs and second order polynomial models. This has led some companies to develop processes based on the application of more sophisticated techniques, including various types of space-filling designs, neural network and stochastic process models, and different global optimisation algorithms.

Advanced modelling techniques are essential for problems where classical DoE is inadequate because of the presence of higher order variable effects and interactions. The advantages and disadvantages of these methods are presented and three examples of engine development activities involving advanced models are described.

The examples, covering cylinder block design, evaluation of new fuel injection equipment and common rail diesel engine calibration, demonstrate that advanced models can be incorporated successfully in the engine development process and deliver improvements in quality through accurate emulation of non-linear variable effects.

1 INTRODUCTION

As the engineering challenges for the automotive industry become more demanding, Design of Experiments (DoE) is increasingly being used in engine development. For applications such as common rail diesel engine calibration, it is necessary to look at alternatives to classical DoE as well as new developments in the field of optimisation.

These alternatives to classical DoE must meet certain requirements if they are to be of practical benefit in the development of engines. The experiment designs (test matrices) must allow particular points to be included, or specific operating regions to be excluded. The modelling technique must not require an excessive number of data points and be able to differentiate between noise and genuine higher order effects. The model generation and analysis times must be short (less than one hour and ideally much quicker) so that time is never wasted on the testbed waiting for results. Optimisation must be fast and provide a high level of flexibility in terms of the number of objectives and constraints that can be handled without compromising efficiency.

2 DOE AND ENGINE DEVELOPMENT

Design of Experiments is now routinely used for engine development by the major manufacturers and leading automotive consultants. This is because modern engines are increasingly sophisticated and therefore demand greater development and calibration effort. For some applications, it is now impossible to conduct a thorough optimisation of the engine and ECU variables in a realistic timescale without using DoE.

Polynomial equations, usually second order, remain the most commonly applied DoE model in automotive engineering. Applications range from research through design, CAE, development, calibration and OBD, to production [1]. Second order polynomial models have significant limitations, but are nonetheless suitable for many engineering problems, especially with thoughtful use of data transformations and some restrictions on the ranges of certain variables [2, 3].

For highly non-linear systems, low order polynomials cannot adequately model response behaviour, and increasing the order of the equation (i.e. including cubic and higher terms) can cause undesirable interpolation effects, especially for multi-dimensional problems. Therefore, it is necessary to consider more sophisticated modelling techniques if DoE is to be applied to multi-dimensional non-linear systems. There are a wide variety of advanced models to choose from, each offering contrasting levels of non-linear capability, statistical information and computational requirement. Examples of these techniques include neural networks (NN), stochastic process models (SPM), generalised additive models, multi-variate adaptive linear splines and piecewise linear models [4].

Advanced models also provide other advantages. In classical DoE, selecting the right ranges for each variable is an extremely important task. If the ranges are too narrow, the optimum may be missed. If they are too wide there may be points in the design matrix that cannot be achieved or where the engine operation is unstable. When this happens the usual requirement that the responses can be approximated with a second order polynomial may not hold and the model will be poor. Since range setting is so important, considerable engineering effort, and sometimes test time, is devoted to selecting correct ranges.

Advanced models, which can handle a sharp upturn in response near the boundary of the design space, or some other non-linear effect, avoid the need to select ranges precisely and reduce the amount of engineering effort and/or testbed time required to define the experiment.

In the field of computational optimisation, gradient (hill-climbing) methods dominate, but more robust global techniques including genetic algorithms, simulated annealing and multi-level co-ordinate search methods are useful for niche problems. Of particular interest to automotive engineers are methods for generating Pareto optimal curves that show the trade-off between different engineering objectives.

3 EXAMPLE APPLICATIONS OF ADVANCED MODELLING TECHNIQUES

3.1 Design of Aluminium Cylinder Blocks for Diesel Engines

Small engine package requirements are at odds with the ever increasing trend for higher specific ratings resulting in hotter, more highly distorted and stressed cylinder blocks. There is a need to understand the limiting effects of key cylinder block physical dimensions and their interactions. This understanding allows an optimum design solution to be found within the particular design limitations imposed.

The objective of the investigation reported here was to assess the effect on cylinder block temperature of varying a range of physical dimensions. The temperature field and the key physical dimensions of the cylinder block have an effect on stress, distortion and durability of the cylinder block. Thus the effect on these responses were also assessed in terms of the key physical dimensions in this Computer Aided Engineering (CAE) study.

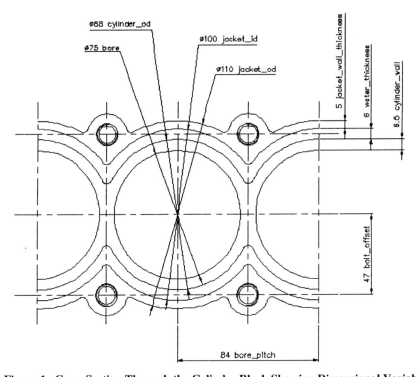

Figure 1. Cross Section Through the Cylinder Block Showing Dimensional Variables

DoE was chosen for this task because it provides:

- Desktop models of the key responses for visualisation and learning about the effect of different design parameters
- A model that enables quicker optimisation of the design because it runs in milliseconds not hours.

A parametric Finite Element Analysis (FEA) model comprising the cylinder block, cylinder head, gasket and head bolts was created in IDEAS [5]. The block was aluminium with parent metal bores. It was assumed that the wear-resistant bore coating would have negligible impact on cooling and it was not included in the FEA model. The base design dimensions are given in Table 1. The design variables of interest and expected limits for the investigation are given in Table 2. These variables, and other dimensions are shown in Figure 1.

A further variable, Specific Rating (kW/l), was added to the experiment. By including the power rating of the engine, a much wider range of engine specifications could be examined with a single model.

Parameter	Dimension (mm)
Bore pitch	84
Bolt offset	47
Bore	75
Cylinder OD	88
Cylinder wall	6.5
Jacket ID	100
Water thickness	6
Jacket OD	110
Jacket wall thickness	5
Cylinder lower OD	93
Block height	222
Jacket depth	67
Counterbore depth	59

Table 1. Basic Design Dimensions for the Cylinder Block Study

Parameter	Baseline	Min.	Max.
Bore pitch (mm)	84	79	87
Bolt offset (mm)	47	44	52
Cylinder wall (mm)	6.5	5	9
Jacket depth (mm)	67	30	100
Counterbore depth (mm)	59	20	120
Specific Rating (kW/l)	60	50	70

Table 2. Experiment Variables & Ranges for the Cylinder Block Study

C606/028/2002 © IMechE 2002

Response	Units
Interbore Peak Temperature	C
Surface Peak Temperature	C
Circ. Range Peak Temperature	C
Circ. Range Average Temperature	C
Bore Extreme Range Temperature	C
Peak Stress	MPa
Yield Levels	Area Assessment
Minimum Durability	Safety Factor
Ring Conformability	Dimensionless Rating

Table 3. Responses Modelled for the Cylinder Block Study

A Latin Hypercube design was chosen to enable a Stochastic Process Model to be fitted. The design selected was an orthogonal one to enable polynomial models to be fitted as well for comparison/validation.

FEA models were generated for each point of the test matrix and used to predict the resultant temperature field. Boundary conditions were applied to simulate thermal loading for the relevant specific rating. Cooling rates were obtained from a database of empirical results. For the DoE modelling, the temperature field results were distilled into summary responses, such as peak temperature or temperature ranges, that are key to the design process. These and the other responses modelled are shown in Table 3.

With noise-free data, SPMs are equivalent to the statistical methods called kriging and DACE [6] and are essentially interpolation between points. Hence, all models had a coefficient of regression, $R^2=1$ (i.e. zero residual error). Any modelling effort is directed towards ensuring that the interpolation is smooth and well-controlled (e.g. by choice of correlation function or data transformations) as the "fit" to the data is always perfect.

Figure 2 shows an overview of the response of Interbore Peak Temperature to all six variables. The dotted lines denote the prediction 95% confidence interval. This view is useful for making an initial assessment of the model. In this case, the variables have the anticipated effects.

Although the response of Interbore Peak Temperature is not especially non-linear, the effect of increasing Jacket Depth has an exponential decay profile (i.e. it slopes downwards then levels out). This shape (see the fourth sub-plot in Figure 1) is one that is difficult to model with a polynomial because the quadratic term tends to make the right hand side "curl up". This feature is illustrated by the equivalent plot from the built-in Matlab™ function *rstool* (shown in Figure 3) for the full quadratic model. The polynomial model also appears to indicate the presence of an effect due to Bolt Offset. However, this is known to have minimal effect on the Interbore Peak Temperature (it was included because of its influence on other responses) and the effect observed in this model could be due to a higher order interaction that is aliased with the quadratic term for Bolt Offset.

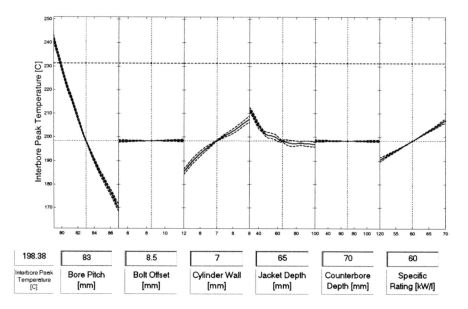

Figure 2 – Overview of Variable Effect on Interbore Peak Temperature
(Stochastic Process Model)

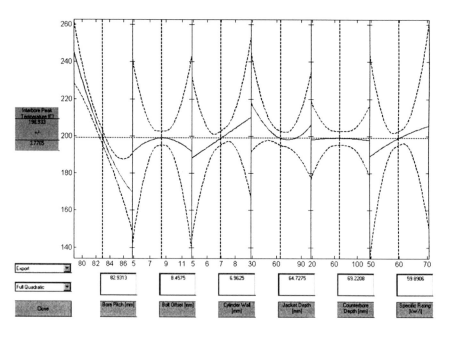

Figure 3 – Overview of Variable Effect on Interbore Peak Temperature
(Polynomial Model)

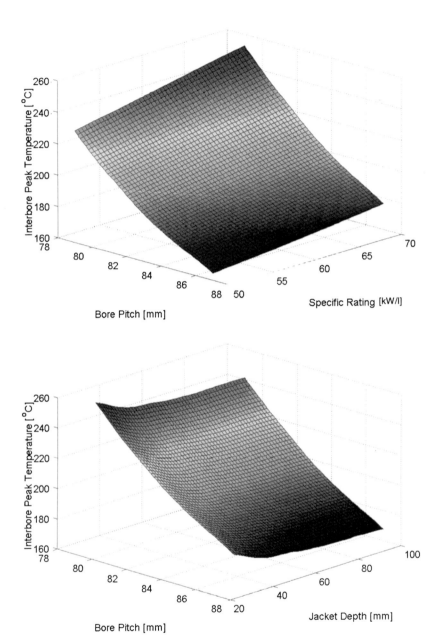

**Figure 4 – Selected Response Surface Plots
(Stochastic Process Model)**

Both the problems with the polynomial model described above may be partially alleviated by a response transformation and stepwise regression (selective removal of coefficients in the model) but, at least in this example, the SPM model produced an excellent fit first time and required no further refinement by the engineer or DoE specialist. This is often the case with noise-free data. Therefore, in addition to a saving in engineering effort where variable range setting is concerned, there is often less work required to improve the model.

The confidence intervals for the polynomial model are much wider than for the SPM. As the data is noise-free, this is due to lack of fit (i.e. the model is not flexible enough to go through each point and residuals are non-zero). If the experiment design had been optimised for fitting a polynomial model, rather than a compromise between two model types, this effect would be less dramatic but still evident.

The surface model plots in Figure 4 show the interaction of the input parameters and the effect on the Interbore Peak Temperature. Bore Pitch is the most significant parameter. The effect of Specific Rating increases with the reduction in Bore Pitch. This is a result of the changing interbore heat path length as well as its width as the Bore Pitch reduces. The effect of Jacket Depth is independent of Bore Pitch and ceases to have an effect on Interbore Peak Temperature over a certain depth (approximately 60mm). This is a result of the relatively high heat input towards the top of the bore compared to the low heat input towards the middle and lower areas. The Bolt Offset and Counterbore Depth have insignificant effects on the Interbore Peak Temperature response (thus are not shown) but do play a significant role in bore distortion levels.

This response surface model can be interrogated with an optimisation algorithm to minimise Interbore Peak Temperature. However, there generally is more than one objective and a number of geometric constraints to comply with. In such cases, some or all the models are used to explore the trade-offs between peak temperature, temperature range, ring conformability (bore distortion) and durability. The Pareto analysis discussed in sections 3.2 and 3.3 works well with this type of problem.

3.2 Evaluation of a New Electronic Unit Injector

A test programme was devised using DoE methods to evaluate the performance of a heavy duty single cylinder diesel engine equipped with cooled exhaust gas recirculation (EGR) and a novel two valve electronically controlled unit injector (EUI). In addition to the normal EUI spill valve, the new injector has a second valve to control needle opening and closing which enables multiple fuel injections and some pressure control between injections within an existing EUI engine architecture.

The objective was to optimise the exhaust gas recirculation level and injection parameters towards achieving legislated exhaust emissions with minimum fuel consumption. Tests were carried out at a single medium speed, ~50% load running condition. The injection variables were timing, initial injection rate as set by the nozzle opening pressure, pilot injection and post injection.

The programme was divided into three stages, looking initially at a single injection with variable nozzle opening pressure (NOP), then at pilot injection and finally with post injection added. For the multiple injection (using pilot and/or post injection) experiment designs both injection duration and separation were included as variables.

Engine responses to secondary injections can sometimes be highly non-linear and therefore not always suitable for analysis by classical DoE. Furthermore, the novelty of the fuel injection equipment meant that the response characteristics were potentially outside existing experience. Hence it would have been extremely difficult to specify the narrow variable ranges necessary for a successful classical DoE while being confident of capturing the optimum. Thus, the problem was ideally suited to an advanced modelling technique, such as stochastic process models, which are capable of modelling non-linear responses over wide variable ranges.

Test plans were created using space filling designs with variable ranges chosen based on previous experience and fuel injector simulations. There existed a number of variable combinations that either could not be achieved by the hardware or which were known to be undesirable. This issue was resolved by excluding unwanted regions of the design space and redistributing points within the desired test space.

Once the tests were complete, models were made for each stage. Model quality was assessed with metrics equivalent to those used with polynomial models and led to the detection of some genuine outliers, which were confirmed by reference to the raw data. Figure 5 shows an example response surface plot from the resulting models.

A viewing function enabled engineers to interrogate the models and gain understanding of the significant variables and interactions for this novel injector. Figure 6 shows an example of the viewer applied to the post injection model.

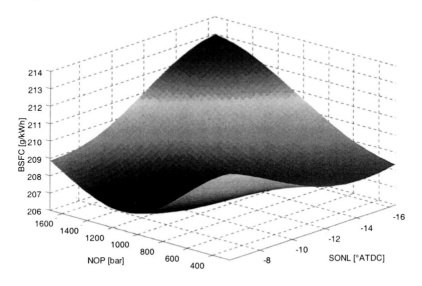

Figure 5 – Example of Fuel Consumption Model Response Surface (Stochastic Process Model)

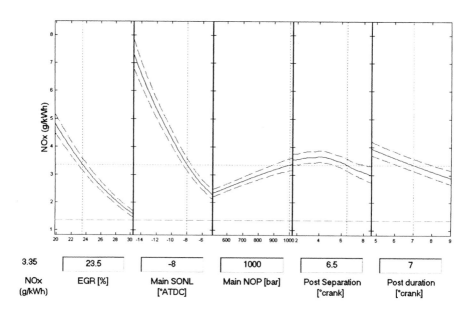

| 3.35 | 23.5 | -8 | 1000 | 6.5 | 7 |
| NOx (g/kWh) | EGR [%] | Main SONL [°ATDC] | Main NOP [bar] | Post Separation [°crank] | Post duration [°crank] |

Figure 6 – Example of Model Viewer for NOx Response with Post Injection (Stochastic Process Model)

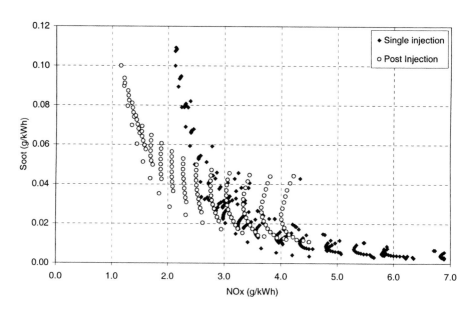

Figure 7 – Side View of Pareto Surface Comparing Single and Post Injection

C606/028/2002 © IMechE 2002

The models were each optimised to produce a 3-way Pareto surface showing the optimum trade-off between NOx, soot and fuel consumption. Best soot and best fuel consumption solutions, at a given NOx level, were thus produced in one step.

Pareto optimisation results for each model were then used to compare the relative merits of high pressure single injection, pilot injection and post injection. Figure 7 is a 2-D view of the 3-way Pareto surface which shows that post injection can provide a significant soot reduction at low NOx but leads to slightly increased soot at higher NOx levels.

The findings of the test programme showed that the new two-valve injector could provide significant soot reductions through increased pressures and multiple injections. The use of DoE techniques was helpful, firstly to gain understanding of a complex system and, secondly to provide information throughout the entire tested space which could be compared between models of the different injection strategies.

3.3 Diesel Engine Calibration

Common rail diesel engines introduced a new level of sophistication to the engine control system, beyond that of gasoline engines (even direct injection gasoline engines). As well as the additional complexity, there is considerable pressure to reduce development time and meet a number of conflicting objectives, namely fuel economy, emissions, driveability and refinement.

In a recent research programme [7] an emissions and performance demonstrator vehicle was calibrated using DoE methods. The demonstrator is powered by a 1.2 litre (70 mm bore by 78 mm stroke), four cylinder light duty common rail diesel engine with a rated power of 75 kW. The HSDI engine utilises Bosch common rail fuel injection equipment developing 1600 bar peak rail pressure and a Garrett variable geometry turbocharger (VGT). Low speed torque is supplemented by an integrated starter generator (ISG). The aftertreatment system consists of a passive deNOx catalyst and a Diesel Particulate Filter (DPF).

As is common practice in diesel engine calibration, the engine is first optimised at a number of "key points", fixed speed load conditions. The key points and their relative importance were determined by simulating the vehicle over the emissions drive cycle utilising the Ricardo vehicle simulation program, VSIM, which models the complete powertrain including the aftertreatment [8]. The engine ECU variables involved in the calibration were standard: injection timing, pilot injection timing, EGR rate, rail pressure and boost pressure. Pilot quantity was fixed at the minimum stable setting.

A 45 case 5 variable orthogonal Latin Hypercube (LHC) was generated for twelve key points. For key points where the boost pressure was not an input variable (VGT position fixed), a 28 case 4 variable orthogonal LHC was constructed.

Figure 8 shows example response plots at the 2600 rev/min, 11 bar BMEP key point. All models were satisfactory at this condition. The gentle undulations in responses against pilot separation, evident even over the small crank angle range tested in this particular experiment, are modelled well by the stochastic process models.

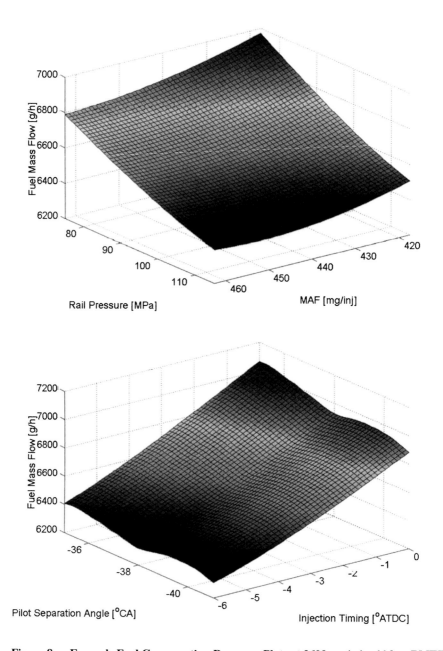

Figure 8a – Example Fuel Consumption Response Plots at 2600 rev/min, 11 bar BMEP

C606/028/2002 © IMechE 2002

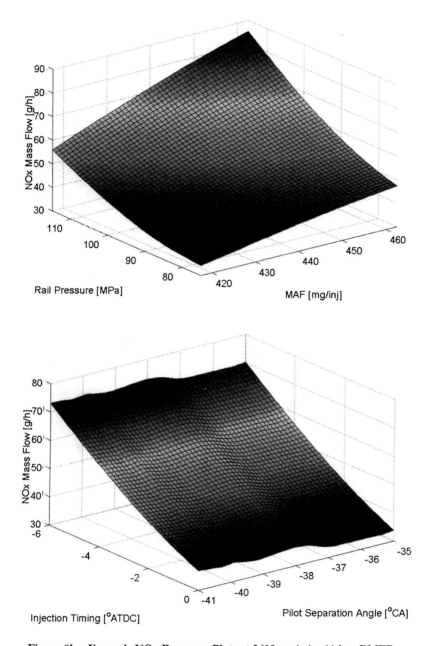

Figure 8b – Example NOx Response Plots at 2600 rev/min, 11 bar BMEP

Figure 9 shows a 3-way Pareto optimal surface for a build of the demonstrator engine over the "Euro 4" emissions test cycle. This surface was generated using a Matlab based in-house software program. With this tool, it is possible to reduce the time taken to generate a 3-way (NOx, PM, fuel consumption) Pareto optimal surface for an emissions cycle represented by twelve key points (with four or five input variables at each point, about 50 variables in total) to less than one hour on a basic desktop PC.

Any position on the Pareto surface represents a particular ECU calibration giving Pareto optimal NOx, PM and fuel consumption. All solutions meet engineer-defined constraints for HC, CO, noise and maximum rate of change in variable settings. The most efficient calibration meeting emissions targets can be selected from this surface. Using this technique the testbed development programme has successfully demonstrated the potential of the demonstrator powertrain to achieve half of the Euro 4 legislated emissions levels and 4 litres/100 km when applied to a C-class vehicle of 1130 kg.

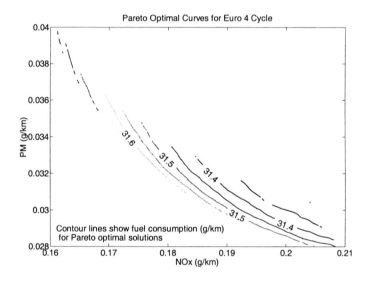

**Figure 9 - Pareto Curve Showing Fuel Consumption Contours
Over an Engine Out NOx versus PM Trade Off
Compatible with Euro 4 Tailpipe Emissions Legislation**

5 CONCLUSIONS

The increasing sophistication of engines and the pressure on development time and quality, have led most manufacturers to utilise Design of Experiments in the engine development process.

While polynomial models work well for the majority of applications, for some development activities, advanced models are preferable – if not essential. The three applications described in this paper, are examples where stochastic process models make the engine development process faster or better in terms of the quality of the solution obtained. The increase in speed comes from not having to conduct preliminary tests or expend engineering effort in defining variable ranges for the experiment. The improvement in quality arises from the better modelling of non-linear effects and the capability for higher dimensional problems.

In diesel engine calibration, emissions cycle optimisation problems may effectively involve fifty or more variables and three or more objectives. Therefore, methods for efficiently generating Pareto Optimal surfaces are essential. The research reported briefly here, has utilised a new Matlab based optimisation tool that significantly reduces the time required to generate such surfaces.

6 REFERENCES

1. Edwards S P, Grove D M & Wynn H P (2000), Statistics for Engine Optimization, PEP, UK, ISBN 1 86058 201 X.
2. Kuder J & Kruse T (2000), "Optimizing Parameters for Gasoline Direct Injection Engines", MTZ Worldwide, pp 8-11 & pp 378-384.
3. Montgomery D T & Reitz R D (2001), "Effects of Multiple Injections and Flexible Control of Boost and EGR on Emissions and Fuel Consumption of a Heavy-Duty Diesel Engine", SAE 2001-01-0195.
4. Holmes C C & Mallick B K (2001), "Bayesian Regression with Multivariate Linear Splines", Journal of the Royal Statistical Society: Series B, Vol. 63, No. 1, pp 3-17.
5. IDEAS - http://www.sdrc.com/ideas/
6. Sacks J, Welch W J, Mitchell T J & Wynn H P (1989b), "Design and Analysis of Computer Experiments", Statistical Science, Vol. 4, No. 4, pp 409-435, 1989
7. Gordon, R L, "A Systems Approach for the Mild Hybrid Powertrain", 2nd Ricardo International Conference – Vehicle Systems in the Wired World, 12th & 13th June, 2001, UK.
8. Goodfellow C L, Burton D J, Fussey P M & Wray J B, "Advanced Integrated Powertrain Modeling", 2nd Ricardo International Conference – Vehicle Systems Integration in the Wired World, 12th & 13th June, 2001, UK.

7 ACKNOWLEDGMENTS

The authors wish to thank the Directors of Ricardo for permission to publish this paper.

8 CONTACT INFORMATION

The authors can be contacted at the following email addresses:

jseabrook@ricardo.com   ghfarrow@ricardo.com   spedwards@ricardo.com.
bjrogers@ricardo.com   japatterson@ricardo.com

Further information on Ricardo DoE techniques can be found at:
http://www.ricardo.com/pages/techdoe.htm .

9 ABBREVIATIONS

BSFC	Brake Specific Fuel Consumption
CAE	Computer Aided Engineering
CO	Carbon monoxide emissions
DoE	Design of Experiments
DPF	Diesel Particulate Filter
ECU	Electronic Control Unit
EGR	Exhaust Gas Recirculation (rate)
EUI	Electronic Unit Injector
FEA	Finite Element Analysis
HC	Hydrocarbon emissions
HSDI	High Speed Direct Injection
ISG	Integrated Starter Generator
LHC	Latin Hypercube experimental design
MAF	Mass Air Flowrate
NOP	Nozzle Opening Pressure
NOx	Nitrogen Oxides emissions
OBD	On-Board Diagnostics
PM	Particulate Mass emissions
SONL	Start of Needle Lift
SPM	Stochastic Process Model
VGT	Variable Geometry Turbocharger

Application of adaptive online DoE techniques for engine ECU calibration

Y RAYNAUD and **D LOPES**
PSA Peugeot Citroën, Paris, France
T FORTUNA, A ZRIM, and **A M GALLACHER**
AVL List GmbH, Graz, Austria

This paper presents an efficient DoE method for planning experiments for ECU calibration. The successful combination of statistical techniques with geometrical concepts guarantees the identification of an optimal design for strongly non-linear systems defined in irregularly shaped experimental spaces, which are not given a priori.

As a main advantage for the calibration engineer, by using the adaptive online DoE the complete procedure on the test bench has been automated requiring almost no pre-investigations and no specific statistical knowledge.

Keywords: EMS & ECU calibration, optimisation, online design of experiments, modelling.

INTRODUCTION

The ever-increasing costs associated with performing experiments at the testbed are a primary problem for the calibration of engines, the other problem to deal with is the ever increasing number of ECU (Electronic Central Unit) parameters to tune to find an optimum calibration. An intelligent way to plan testing includes the use of Design of Experiments (DoE). DoE allows identifying the small number of tests that provide the greatest amount of information on the system and the experimental space in order to reduce calibration time and cost by reducing testbed and prototype engine running time.

THE ENGINE CALIBRATION TASKS

The use of DoE techniques for calibration engines problems is not straightforward : the big difficulty lies in the fact that the boundaries of the valid experimental space are usually unknown at the beginning of the investigation into new engine hardware and ECU strategies. Usually, the unknown irregular shape of the design space makes the task of planning tests particularly difficult. The operating ranges of the engines are often restricted by limitations on injection systems and on further parameters such as EGR, boost pressure. The identification of the feasible boundaries is a time consuming task for the engineers requiring several manual calibrations and human interpretations. Moreover, deriving a mathematical formulation of the boundaries from such investigations is almost impossible for problems with many variation parameters.

THE INNOVATIVE APPROACH FOR ENGINES ECU CALIBRATION

DoE is widely and successfully used in industrial and academic fields. The first experiments with a mathematically planned structure were designed by R. A. Fisher at the Rothamsted agricultural research station in the 1920's. In the 1940's the theory of fractional factorial designs was developed by Fisher and others. In the same period, the clear link between DoE, statistical modeling and optimization was observed for the first time. In the 1950's G. Taguchi generalized the theory of Fisher and applied this to technical areas. A new type of experimental design, called Optimal Design, emerged from the research of J. C. Kiefer in the 1950's. Kiefer introduced the concept that different designs for an experiment could be evaluated by measuring the worth of each design. In this way it would be possible to identify the design that is *optimal* by this measure. More details can be found in [1, 2].

The application of statistical experimental design in the automotive field is not new. Some previous publications on this topic can be found in [3, 4, 5, 6, 7, 8].

Our work distinguishes itself from previous publications by the ability to automatically identify a mathematical approximation of the feasible design space and at the same time, an optimal design inside it. Aware of the difficulties in planning experiments for ECU calibration, we have developed a fully automatic procedure that identifies the feasible design space on-line and, at the same time, generates a D-optimal design inside the feasible region. The method is based on a refinement approach, which adapts on-line the design to the irregularly shaped range.

The procedure does not require any additional offline interpretation which is usually time consuming. Due to its strong mathematical fundaments, the method is robust and applicable for high dimensional and nonlinear constrained problems.

ADAPTIVE ONLINE DOE APPROACH

The method generates in the first stage a preliminary design (Full Factorial, Central Composite, or D-Optimal) in the given variation range. Starting from a central stable point, the procedure tries to reach stepwise the points of the design, checking at each step the validity of the restrictions. As soon as a limit violation occurs, the procedure measures at the previous stable location. The aim of this first stage is a full investigation of the design space in order to detect points lying on or close to the boundaries of the valid region. Figure 1 shows the first stage performed with a Central Composite Faced Design for a two-dimensional design space. The method guarantees as many measured points, as the specified design requires.

C606/021/2002 © IMechE 2002

Due to the limit violations, the statistical correctness of the design is no longer guaranteed. In order to ensure a high quality design, and thus a model with a small prediction error and small confidence intervals, the goodness that can be reached with the measured points is evaluated. If the quality is not satisfactory, the procedure computes a mathematical approximation of the valid experimental space (Figure 2) and, by means of an optimisation algorithm, adapts the design by iteratively adding some additional points in the valid range until a certain quality level is reached (Figure 3). These pictures with more details can be found in [9].

The final result is a D-optimal design maximising the information on the model to predict.

The simple two-dimensional example here illustrated helps to easily describe the basic principles, but the procedure works successfully in multidimensional spaces as well.

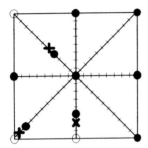 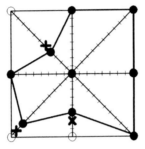 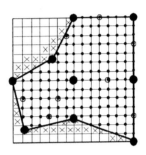

Figure 1. At the first stage, the procedure performs a full investigation of the valid region, detecting points lying close to the boundaries. In this case the method has tried to reach the points of a Central Composite Faced Design. The test points are denoted by black dots. A red cross indicates a limit violation.

Figure 2. Based on the previously measured points, a mathematical approximation of the valid design region is computed.

Figure 3 A refinement of the previous design is done. Red circles denote additional points improving the quality of the original design (denoted by the back dots).

APPLICATION OF THE THEORY

The Adaptive D-Optimal design method has been used to calibrate a HSDI common rail engine. The target was to minimize the fuel consumption, combustion noise while fulfilling euro 4 exhaust gas emissions limitations.
In order to achieve the task, the following five ECU parameters were screened:
- Start of Pilot Injection
- Pilot Injection quantity
- Start of Main Injection
- Rail Pressure
- EGR-rate

During the execution of the test, several physical quantities have been measured, in particular fuel consumption, smoke, CO, NOx, HC, stability of combustion (by means of IMEP stability).

At PSA Peugeot Citroën the main idea is to model the ECE and EUDC dynamic test by several stationary key points (i.e. stationary speed / load reference points) with weighing factors. This approach allows us to get a first ECU calibration in classical stationary testbeds before the first car prototype is available. The draw back of this method is that key points are specific to the car inertia and the gearbox ratio. Furthermore the same engine could be applied to different cars or gearboxes ratio. To work around this draw back and to be able to find several optima for different car application the idea was to generate a DoE that takes into account engine speed and load variation in order to create an extended model. In this case several numerical optimisations on the same model are run to find different optima depending on the car application. The speed and load area covered during this study was between 1600 – 2200 rpm and 50 to 100 Nm on a 2 litre HDI engine. Varying such speed and load sites with in addition variations in rail pressure, start of injections and pilot quantity leads to important variations on the exhaust temperature and then the energy available for boost and EGR capability.

The calibration engineer is not able to predict which EGR rates are feasible or not and for that reason the DoE online adaptive function was used in CAMEO to identify differences between the designed EGR rate demands and the real EGR ratio that the engine can achieve. That functionality automatically readapts the EGR demand in order to create a good model for the engine responses on this parameter.

A first design of experiment is proposed to CAMEO in order to include an existing engine constraint and avoid a waste of time for this particular constraint. The design should be suitable to fit a second order model of the consumption and emission responses. Since the 7-variable experimental domain is constrained, a D-Optimal design with 79 experimental units was built (73 points + 6 center points). D-optimal designs reach to spread out the experimental units over the feasible domain. Mathematically, it consists in maximizing the det(X'X), where X is the orthogonalized experimental matrix. A KL-Exchange algorithm developed by Federov and presented in [2] was used to provide the best matrix. Computing time was less than 5 minutes to extract a 73-points design from a set of 643 candidates. The procedure was repeated several times as the Federov algorithm is influenced by the random-choice of the starting design. The second phase was planned in order to model polynomial models of the second order and obtain a better prediction of the engine responses leads by EGR rate.

Figure 4 shows the measured points in a 3D projection. The vectors coming out from the central point display the stepwise move of the strategy till the planned points. The limit violations are displayed as cross. As a reaction to limit violations the method stepped back and measured in correspondence of the last stable point identified on that vector.

C606/021/2002 © IMechE 2002

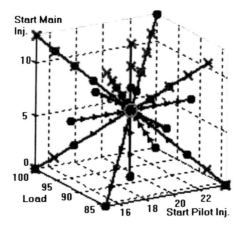

Figure 4. Limits encountered between EGR demands in the design and EGR ratio really achieved.

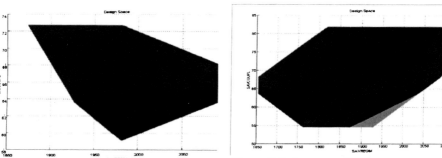

Figure 5. Rail pressure 450 bar.

Figure 6. Rail pressure 550 bar.

Figure 5 displays a feasible region for EGR rate according to speed and load variations for 450 bar of rail pressure cutting plan identified after the first phase. The boundary of the design space is characterised by several facets, which would be difficult to detect by means of manual calibrations.

Figure 6 displays the feasible region for EGR rate according to speed and load variations for 550 bar cutting plan. Figure 7 displays the same feasible region for EGR rate according to speed and load variations for 450 bar cutting plan but with a different cut in pilot advance The different colours, which can be seen in Figure 5, 6 and 7, refer to different levels of EGR-Rate, while the other variations are kept constant to certain values.

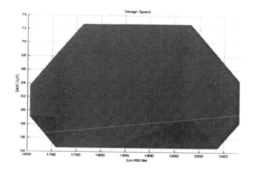

Figure 7. Rail pressure 550 bar with another cut in start of pilot injection.

After the identification of these limits, the system is then online calculating the best D-Optimal plan to fit inside the limitation border to achieve the same level of confidence in terms of number of measurements and space between points as in the original design that didn´t take into account the EGR borders (Figure 8). These new points are then sent to the ECU and measured automatically by CAMEO.

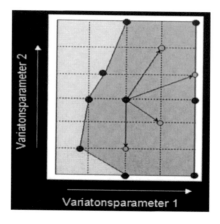

Figure 8. Border limitation and D-optimal plan redesign.

MODELLING AND STATISTICAL ANALYSIS OF THE DATA

A first descriptive analysis of the raw data only reveals a few outliers. The data obtained from the testbed has been modelled by means of advanced polynomial models (second order model with a backward selection i.e. the system automatically removes statistical non significant terms from the model) and FNN (Fast Neural Networks). FNN are mathematical models obtained as weitghed sum of polynomial models.

Local polynomial models are generated in different ranges and they are combined to a global, in general nonlinear, model valid in the whole operating region. The first phase of the model generation is to fit a second order polynomial model on each response. The advanced model function automatically looks for a transformation and adjusts the number of terms of the polynomial. Since some relations are more complex, FNN modelling is preferred. For example, the advantage of FNN modelling in figure 9 and 10 can be seen. One of the main complex features is the big increase in emissions that is not predictable by usual quadratic models. For the fast neural network model procedure the system splits the domain and fits a polynomial model on each zone. The choice of the best split strategy is automatic.

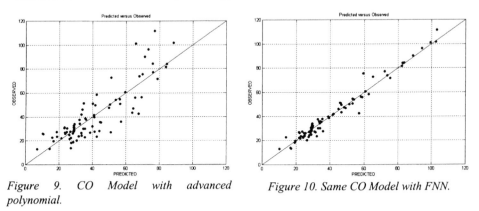

Figure 9. CO Model with advanced polynomial.

Figure 10. Same CO Model with FNN.

Nevertheless, it is useful to check the results. Attention will be on R-square, outlier analysis, prediction standard deviation and residual analysis Moreover, the shape of engine responses should be validated by the calibration engineer.

The responses are BSFC, combustion noise, emissions and several combinations of measurements. These combinations are called 'Target Function' and will allow us to minimize multi-criteria objectives while constraining emissions. This target function is adimensional and mainly based on BSFC and combustion noise (see Figure 11).

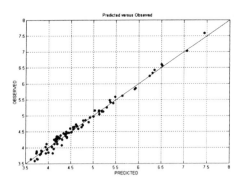

Figure 11. Target function model.

Since CAMEO fits user-specified target function, one will need to have a good prediction of this quantity. Figure 11 shows the prediction vs the observed values for the target function.

Figure 12 shows the influence of each parameter variation on BSFC. The dotted line is the prediction 95% confidence interval.

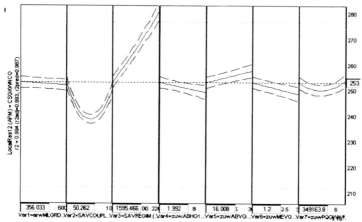

Figure 12. BSFC Model R² = 0.994 R² pred = 0.98.

OPTIMISATION OF THE CALIBRATION

From the global speed and load models generated before the CAMEO optimisation toolbox was used in order to minimize fuel economy and noise within constraints on the exhaust emissions. In order to compare the new predicted calibration sets to the actual ECU settings the system looks for the best feasible parameter sets for two speed and load points.

1700 rpm 78 Nm				2150 rpm 76 Nm		
	Target	Optimum			Target	Optimum
BSFC		228	BSFC			231
Noise (dB)	< 86	86	Noise (dB)		< 84	83
CO (g/h)	20.4	15.6	CO (g/h)		23.2	22.6
HC (g/h)	3.2	2.4	HC (g/h)		4.21	4.2
NOx (g/h)	38.5	33.5	NOx (g/h)		41.2	41.2
Smoke	< 3	2.3	Smoke		< 3	2.5

The target was implemented in CAMEO by a numerical function to minimise which was a non dimensional combination of BSFC and Noise, all the engines responses including exhaust emissions, running stability, maximum combustion noise and maximum fuel consumption were set as local constraint, no solution would be proposed above these values.

We should also assume that the optimization is very much dependent on the weighing function coefficients. This is still the key role of the calibration engineer to decide if he would rather diminish the noise but with higher exhaust emissions. This compromise remains under the calibration engineer responsibility.

What we should also mentioned is that the test bed was used during 12 hours in order to run the DoE completely, then only half a day was necessary to validate the model and begin to find out optima settings for the ECU, note that this work is of course done out of the testbed with the offline version of CAMEO.

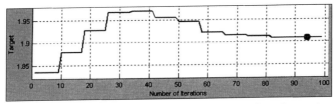

Figure 13. Automatic Optimisation Process with sequential quadratic algorithm.

CONCLUSIONS

This paper presents an efficient method of planning online experiments for irregularly shaped design spaces, whose boundaries are unknown a priori. In particular, the procedure has been successfully applied for the calibration of internal combustion engines. The well-known advantages of classical DoE approaches have been combined with new statistical and geometrical ideas. As a result, the method is able to decrease testing times and costs and to greatly reduce the engineers' effort in identifying the feasible experimental region. Moreover, it delivers a precise mathematical formulation of such regions even including speed and load variation, useful for the search of a feasible optimum. The time needed to generate an accurate model and a new calibration setting is then not only diminished but also achieved with less engineer presence.

REFERENCES

1. Grove D., Statistical Modeling and Design of Experiments, Distance Learning Module, University of Bradford 1998.
2. Atkinson, A.C. and A. N. Donev, Optimum Experimental Designs, Oxford University Press, New York, 1992.
3. Piock W.F., Leithgoeb R., Philipp H., Gschweitl K. and Fraidl G., Applikationsmethodik für neue Ottomotorenkonzepte, 22. Internationales Wiener Motorensymposium, 26-27 April 2001, Vienna.
4. Röpke K., Fischer M., Efficient Layout and Calibration of Variable Valve Trains, SAE Technical Paper Series, SAE 2001 World Congress Detroit, Michigan, March 5-8, 2001.

5. Dingel O., Röpke K., Gumprecht D., Application of Design of Experiments Techniques for a Prototype Engine with the Promo Charge-Cycle Program, Proceedings of the 4th International Symposium for Combustion Diagnostic, 18-19 Mai 2000, Kurhaus Baden-Baden.

6. Kuder J., Kruse T., Systematic method of proceeding to optimize parameters for gasoline direction injection engines, Proceedings of the 4th International Symposium for Combustion Diagnostic, 18-19 Mai 2000, Kurhaus Baden-Baden.

7. Reisenweber K.-U., Mitterer A., Fleischhauer T., Zuber-Goos F., BMW Group München, Model-Based Optimization- a New Approach to Increase Efficiency in Control-Unit Calibration, Proceedings of the 4th International Symposium for Combustion Diagnostic, 18-19 Mai 2000, Kurhaus Baden-Baden.

8. Renninger P., Daudel H., Hohenberg G., The DaimlerChrysler Concept for Computed Aided Engine Optimization, Proceedings of the 4th International Symposium for Combustion Diagnostic, 18-19 Mai 2000, Kurhaus Baden-Baden.

9. Gschweitl K, Pfluegl H., Fortuna T. and Leithgoeb R., Increasing the Efficiency of Model-Based Engine Applications through the Use of CAMEO Online DoE Toolbox, ATZ Automobiltechnische Zeitschrift, No. 7-8/2001.

10. Stuhler H., Kruse T., Stuber A., Gschweitl K., Pfluegl H., Piock W., Lick P., Automated Model-Based GDI Engine Calibration Adaptive Online DoE Approach, SAE 2002.

11. Gschweitl K, Pfluegl H., Fortuna T. and Leithgoeb R., AVL CAMEO Online DoE – Wie verkürzt die neue Toolbox die modellbasierte Motorapplikation?, Haus der Technik E-H030-12-036-1, 6-7.Dezember 2001, Berlin Germany.

12. Koegeler H.-M., Fürhapter A., Mayer M. and Gschweitl K., DGI-Engine Calibration, Using New Methodology with CAMEO, ICE 2001, 23-27 September, 2001, Capri-Italy.

13. Fürhapter A., Piock W. and Leithgoeb R., Optimierung von neuen Brennverfahren am direkteinspritzenden Ottomotor, Tagung Direkteinspritz im Ottomotor, Haus der Technik, 27-28 Jun, 2001, Koeln.

14. Barber, C. B., D. P. Dobkin and H. T. Huhdanpaa, The Quickhull Algorithm for Convex Hulls, *ACM Transactions on* Mathematical Software, Vol. 22, No. 4, Dec. 1996, p. 469-483.

15. Grünbaum B., Measure of symmetry for convex sets, Proceedings of the 7th Symposium in Pure Mathematics of the American Mathematical Society, Symposium on Convexity, AMS, Providence, R.I., 233–270, 1961.

C606/022/2002

Radial basis functions for engine modelling

T M MORTON
Consulting Services, The Mathworks Limited, Cambridge, UK
S KNOTT
Powertrain As-Installed CAE, Ford Motor Company, Dunton, UK

ABSTRACT

The complexity of engines responses and the number of controllable parameters in engines have been increasing dramatically in recent years. Modelling using the underlying physical equations can be time-consuming and is sometimes an over-simplification. Statistical modelling based on measured data is an important alternative technique. The classical approach has been to use polynomials in all input factors, but this can be insufficient to accurately describe complex engine responses. Additional model flexibility is needed. One approach is to use a type of neural network model, called a radial basis function (rbf) network. In this paper, we introduce rbf networks and demonstrate their superiority over polynomials for modelling MBT (minimum spark for best torque) for spark ignition engines with twin independent variable valve timing (TI – VVT). We discuss the need for optimised selection of the non-linear parameters in the rbf network. All results presented are obtained using the Model-Based Calibration Toolbox, available from the MathWorks.

1. OVERVIEW

The automotive industry faces the competing goals of producing better performing vehicles and keeping development time and costs low. It is crucial to be able to produce fuel-economic vehicles that meet both the legislated emission requirements and the high-performance demands of the consumer. In order to deliver such vehicles in a reasonable timescale, companies are making use of models of engine responses that can be used throughout the design process.

There are two main approaches to engine modelling. The first is to model underlying chemical and physical processes occurring in the engine via differential equations. This approach is invaluable in the early stages of engine design before an initial prototype is built and for developing a first-principal understanding of the engine behaviour. However, the solution of the differential equations can be time-consuming, which can mean that the physical models are impractical for multi-objective optimisation routines and hardware-in-the-loop simulations. Moreover, when there are a large number of control parameters and engine responses are complex, then the physical model can be over-idealised and fail to capture all the trends in the 'real' engine system described by measured data. Consequently, the difference between the physical model predictions and the measured data can be prohibitively large for tasks such as calibration. An alternative to physical modelling is to model the empirical data directly using a statistical model.

The application of statistical modelling in the automotive industry is growing rapidly [1]. The emphasis until recently has been on the use of polynomial models, however for some engine responses, low order polynomials do not have sufficient flexibility to describe all the trends in the data over the entire region of operability. Simply increasing the order of the polynomials used can provide additional flexibility, but it is a recognised problem that this can lead to unnatural oscillations in the model response, especially towards the boundary of the operating region. Another approach that offers the necessary increase in flexibility is to use neural networks [2]. In this paper we introduce a particular type of neural network called a radial basis function network, and apply this technique to model MBT (minimum spark advance for best torque) for a twin independent variable valve timing (TI-VVT) engine. A comparison of the radial basis function approach and the polynomial approach is made as applied to this modelling problem.

2. INTRODUCTION TO RADIAL BASIS FUNCTIONS

There are several strands to the history of radial basis function networks. They have been an active topic of interest in the neural network community [3] and in the approximation theory community [4]. There is also some overlap with a technique called Kriging that originates in the geostatistics community [5]. Radial basis functions are a powerful tool that has proved useful for fitting data points that are scattered or have a large number of input parameters.

We consider a general model form

$$\hat{y}(\bar{x}) = \sum_{i=1}^{p} \beta_i f_i(\bar{x}),$$ (1)

where $\bar{x} = (x_1, x_2, ..., x_n)$ is a vector in the n-dimensional factor space, $\beta_i, i = 1, 2, ..., p$ are the coefficients or weights, and $f_i, i = 1, 2, ..., p$ are the terms in the model. This model is linear in the coefficients $\beta_i, i = 1, 2, ..., p$. Both radial basis functions and polynomials take this general form. For example, taking the terms $f_1 = 1, f_2 = x_1, f_3 = x_2, f_4 = x_1^2, f_5 = x_x^2, f_6 = x_1 x_2$ gives quadratic polynomial model $\hat{y}$ with an interaction order of 2. Radial basis function networks are obtained by taking

$$f_i = K\left(\frac{\|\bar{x} - \bar{c}_i\|}{\sigma}\right),$$ (2)

where $\bar{x}=(x_1,x_2,...,x_n)$ is the vector input, $\bar{c}_i=(c_{i1},c_{i2},...,c_{in})$ is the center of the ith radial basis function term, $\|\bar{x}-\bar{c}\|$ denotes the Euclidean distance between the input and a center c, given by

$$\|\bar{x}-\bar{c}\|=\sqrt{(x_1-c_1)^2+(x_2-c_2)^2+...+(x_n-c_n)^2}\,,\qquad\qquad(3)$$

$\sigma>0$ is the called the width of the radial basis function, and K is a univariate function called the radial basis function kernel. Table 1 lists the commonly used kernels.

Table 1 Radial basis function kernels

Name	Kernel
Gaussian	$K(r)=e^{-r^2}$
Multiquadric	$K(r)=\sqrt{r^2+1}$
Inverse multiquadric	$K(r)=1/\sqrt{r^2+1}$
Thin-plate spline	$K(r)=r^2\log r$
Logistic	$K(r)=1/(1+e^r)$

The neural network community has tended to focus on Gaussian or logistic kernels. In the case of Gaussians, the width is the standard deviation. A radial basis function network with 2 input variables can be pictured as a weighted sum of radially symmetric hills or bowls, centered at scattered locations around the input space.

As in the polynomial case, the linear parameters or weights are determined by solving a matrix system that gives the least-squares fit to the data points. In this way the radial basis function training algorithms differ from the majority of neural network training algorithms that are based on a (often time-consuming) gradient-based approach. As is the case with most non-parametric models, the weights do not have a straightforward physical interpretation.

Radial basis functions networks contain a number of non-linear parameters to set or determine, namely the widths and the center locations. We demonstrate the dramatic effect of the choice of these non-linear parameters on the quality of the fit.

We will show that there are benefits of using kernels other than Gaussians, such as thin-plate splines. Thin-plate splines are widely used in the approximation theory community as the 2-dimensional analogue of natural cubic splines. The natural cubic spline is the curve that passes through a set of points $(x_i,y_{(i)})_{i=1,...,N}$ in a way that minimises the 'bending energy' (measured by the second derivative). A thin-plate spline network (supplemented by a polynomial of order one) is the solution to the analogous problem of passing a thin plate through a set of points $(x_i,y_i,z_{(i)})_{i=1,...N}$ in a way that minimises the 'bending energy' of the plate.

3. PERFORMANCE MEASURES

The root mean square error (RMSE) is calculated by

$$RMSE = \sum_{i=1}^{N} \frac{\left(y_{(i)} - \hat{y}(\bar{x}_i)\right)^2}{N-p} \,. \tag{4}$$

This is a good measure of how well our model is fitting the measured data, but it is not necessarily a good measure of the predictive capability of the model. As an example of this, it is possible to make the residuals very small by fitting a model with sufficient flexibility. However, the predictive capability would often be poor, as the flexibility would be used to fit all the noise in the data. In fact, it has been demonstrated that RMSE under predicts the estimated prediction risk. See (6) for example.

An alternative performance measure is needed. The criterion we focus on is PRESS (predicted sum square error), which is defined using a cross-validation technique. Consider taking one data point, $(\bar{x}_i, y_{(i)})$ away from the data set and fitting the model to this new data. This new model can be used to make a prediction $\hat{y}_{(i)}$ of the response at $\bar{x}_i$. The difference between this prediction $\hat{y}_{(i)}$ and the measured value $y_{(i)}$ is calculated. This is repeated for each of the data points, and the differences are squared, summed, and averaged to give the PRESS statistic:

$$PRESS = \sum_{i=1}^{N} \frac{\left(y_{(i)} - \hat{y}_{(i)}\right)^2}{N} \tag{5}$$

Minimising PRESS is consistent with the goal of obtaining a regression model that provides good predictive capability over the experimental factor space, assuming that the experimental design contains sufficient information content to reveal the true underlying nature of the function to be estimated.

The most important performance measure is how well the model predicts the engine behaviour. We validate the engine models by computing the error between the model and the data not used in the model fit.

4. MODELLING MBT IN VVT ENGINES

We will apply radial basis functions to the problem of modelling MBT (minimum spark advance for maximum braking torque) for spark ignition engines with twin independent variable valve timing. MBT is modelled as a function of engine speed (N), load (L, normalised air charge), intake valve opening (IVO), exhaust valve closing (EVC), and air-fuel ratio (AFR). The MBT measurements are taken at a set of 251 operating points in the 5-factor input space. The operating points are generated using a Latin Hypercube, a space-filling design-of-experiment technique (7). The Latin Hypercube technique ensures that if the design size is N, then there are N projections onto each factor axis. Moreover, the points

C606/022/2002 © IMechE 2002

should be well spaced, for example, by maximising the minimum distance between the points. This design allows a large number of potential models to be investigated. We split this data set into two, one set of data to be fitted, and one set of data for validation purposes. These sets have 125 and 126 data points respectively, chosen by taking every other point from the original data set.

Note that a data set of size 125 may be considered on the small side for an engineering application such as the calibration of a powertrain control module. If increased accuracy is required, then the number of observations used should be increased to capture the full complexity of the TI-VVT MBT response, especially in the low speed region. At low speed, gas dynamics, intake and exhaust gas exchange and trapping effects can have highly non-linear effect. However the data sets are of sufficient size to illustrate the difference in performance between radial basis functions and polynomials.

At each operating point the spark advance is swept from a low to a high value to identify maximum brake torque of the engine. This data is then fitted using a segmented polynomial (8). The segmented polynomial includes a parameter that provides a direct estimate of the MBT timing.

A polynomial is fitted to the MBT "data" that is cubic in each of the 5 input factors, with a maximum interaction order of 3. This gives a polynomial with 56 terms. We improve the fit by applying a stepwise procedure that iteratively removes/adds terms that result in an improvement in the PRESS statistic. This results in a polynomial with 23 terms.

We compare the performance of the polynomial with a variety of radial basis function networks. To emphasise the importance of choosing the kernel, width and centers appropriately, initially we use a Gaussian radial basis function with centers chosen randomly from the data set and a width of 1. This gives poor results. We improve on this model by using a center selection algorithm, applying a search algorithm to determine a good width, increasing the number of terms used, and by changing the kernel to a thin-plate spline kernel. The center selection algorithm used is ROLS (regularised orthogonal least squares), where the centers are chosen from a candidate set consisting of all the data points. The centers are selected in a forward selection procedure; at each step the center that reduces a regularised error measure the most is selected (9). The width selection algorithm employed is a guided trial and error approach. More details on these algorithms are available in the documentation for The Model-Based Calibration Toolbox available from The MathWorks (http://www.mathworks.com/products/mbc/). The results are shown in Table 2.

As we would hope, models with a smaller PRESS statistic perform better on the validation data. Moreover, we see that as more of the non-linear parameters of the radial basis function network are optimised, the quality of the model improves. It is interesting to note that, for this data set, it is only when all possible options are used (optimising centers, widths and kernel) that the radial basis function network out performs the polynomial model with stepwise applied. From an engineering perspective, the differences in the predictive quality between the polynomial and the thin-plate spline network are often only significant at the boundaries of the operating region, for example, at high speed.

Table 2 Results of modelling MBT

Model Type	RMSE	PRESS	Validation RMSE
Cubic Polynomial, 56 terms	3.06	6.27	4.78
Cubic Polynomial, stepwise applied, 23 terms	2.94	3.30	4.38
Gaussian kernel, random centers, width = 1, 56 terms	4.95	6.31	6.56
Gaussian kernel, optimised centers, width =1, 32 terms	3.42	3.99	5.23
Gaussian kernel, optimised centers and width, 24 terms	3.02	3.39	4.45
Thin-plate spline, optimised centers and width, 51 terms	1.62	2.05	3.83

It could be argued that in the results above, we have treated the polynomial unjustly by asking it to fit data generated by a space-filling design rather than a design more suited to the polynomial, such as a d-optimal design (10). To investigate this situation, we generate a d-optimal design with 80 data points using the 125 points that were fitted in Table 2 as a candidate set. We compare the quality of fit of a cubic polynomial model (whose terms have been reduced using stepwise) with a radial basis function network whose width, centers, and kernel choice have been optimised in Table 3. Again we see that the radial basis function network out performs the polynomial model.

Table 3 Results using an 80-point data set that is d-optimal for a cubic polynomial

Model Type	RMSE	PRESS	Validation RMSE
Cubic Polynomial, 56 terms	3.47	8.26	5.42
Cubic Polynomial, stepwise applied, 26 terms	2.86	3.47	4.81
Thin plate spline, optimised centers and width, 48 terms	1.32	2.48	4.02

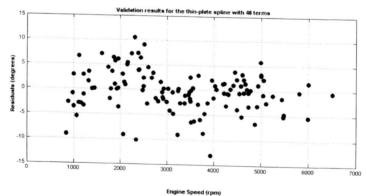

Figure 1 Validation residuals for the thin-plate spline

C606/022/2002 © IMechE 2002

Residual plots on the validation data and response surfaces for the thin-plate spline kernel and the stepwised cubic polynomial are shown in Figures 1-4. Note that the residuals are smaller in the case of the thin-plate spline, especially towards the ends of the range of speed. This illustrates the ability of radial basis function networks to be locally adaptive.

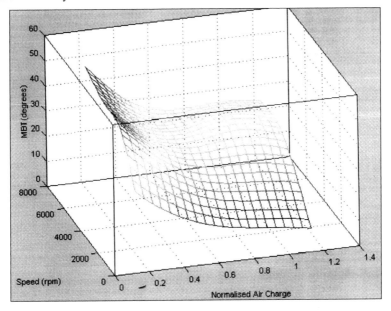

Figure 2 Surface plot of thin-plate spline at IVO = -6, EVC = 6 and AFR = 14.3

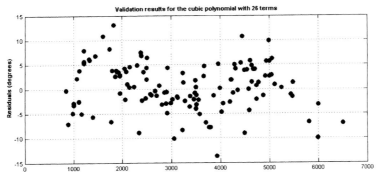

Figure 3 Validation residuals for the cubic polynomial

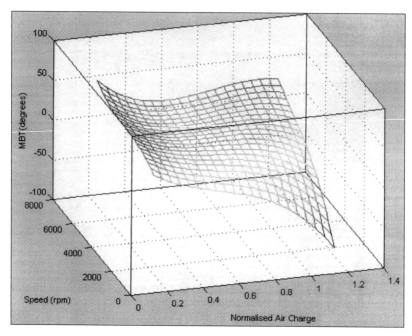

Figure 4 Surface plot of the cubic polynomial at IVO = -6, EVC = 6 and AFR = 14.3

5. CONCLUSIONS

Radial basis functions are an important technique for modelling complex engine responses. Their increased flexibility can capture more of the underlying trends than polynomials across a larger operating region provided that their non-linear parameters are determined carefully.

NOTATION

IVO – Intake valve opening (degrees)
EVC – Exhaust valve closing (degrees)
AFR – Air-fuel ratio

ACKNOWLEDGEMENTS

The authors would like to thank Mark Cary of Ford Motor Company for his valuable input and guidance on the content of the paper.

REFERENCES

1. **Edwards, S. P., D. M. Grove, and H.P. Wynn,** Statistics for Engine Optimization, Professional Engineering Publishing, 2000.
2. **Shayler P. J., Dow P. I., Hayden D. J., and Horn G.**, Using neural networks in the characterization and manipulation of engine data, in Statistics for Engine Optimization, Eds: Edwards, S. P., Grove, D. M., Wynn, H. P., Professional Engineering Publishing, 145—161, 2000
3. **Orr, M.,** Introduction to radial basis function networks, Technical report, Center for Cognitive Sciences, University of Edinburgh, 1996.
4. **Buhmann, M.,** Radial basis functions: the state-of-the-art and new results, Acta Numerica 9 (2000), 1-37.
5. **Cressie, N.,** Kriging nonstationary data, Journal of the American Statistical Association 81, 625-634, 1986.
6. **Eubank, R. L.,** Nonparametric Regression and Spline Smoothing, Marcel Dekker, Second Edition, 1999.
7. **McKay, M., R. Beckman, and W. Conover**, A comparison of three methods for selecting values of input variables in the analysis of output from a computer code, Technometrics 21, 239—246, 1979.
8. **Holliday, T., A. J. Lawrance, and T. P. Davis,** Engine-Mapping Experiments: A Two-Stage Regression Approach, Technometrics, 1998, Vol. 40, pp 120-126.
9. **Chen S., E. S. Chng and K. Alkadhimi,** Regularized Orthogonal Least Squares Algorithm for Constructing Radial Basis Function Networks, Int J. Control, 1996, vol. 64, No, 5, pp 829-837.
10. **Atkinson, A. C. and A. N. Donev,** Optimum Experimental Designs, Oxford University Press, 1992.

C606/025/2002

An engine mapping case study – a two-stage regression approach

D W ROSE and **M CARY**
Ford Motor Company, Laindon, UK
S B ZULCZYK and **R SBASCHNIG**
Ford Motor Company, Detroit, Michigan, USA
K M EBRAHIMI
Department of Mechanical and Medical Engineering, University of Bradford, UK

ABSTRACT

Engine mapping is the process of empirically modelling engine behaviour as a function of the adjustable engine parameters. Engine mapping is empirical in nature and so draws heavily on statistical methods. This paper presents an engine mapping case study conducted on a Ford 1.0L 8-valve engine. The paper concentrates on the modelling of a single response using two-stage regression techniques. This analysis procedure reflects the inherent structure in the data imposed by the experimental method, where data are collected as spark sweeps. At stage-1, curves are fitted to the measured responses as a function of spark advance alone. To improve the engineering interpretation of the modelling procedure, response features summarising the important geometric characteristics of the sweep are defined. Hybrid B-spline-polynomial models are used at stage-2 to explain the variation in the response features with the remaining engine parameters. A computer-generated design of experiment was utilised to ensure efficient recovery of the proposed stage-2 model. Model diagnostics and validation data are presented.

1   INTRODUCTION

Engine mapping is the process of empirically modelling engine behaviour as a function of the adjustable engine parameters [1, 2, 3, 4]. For example, in this study a model is sought that explains the systematic variation in brake torque (T) as a function of spark advance (S), engine speed (N), normalised induced air charge or load (L) and air-to-fuel ratio (A). Although engine mapping models are often embedded in larger dynamic simulations designed to study complex transient behaviour [5, 6], their primary application is the calibration of electronic engine controllers.

The engine controller is connected to various sensors and actuators that allow the state of the power train to be continually monitored and altered when required. A computer program or *strategy*, executed in a continuous loop, determines and sets the desired state. For example, undesirable driveline *clunk* and *shuffle* phenomena [7] can be effectively damped by modulating the spark timing during rapid throttle transients [8]. During such manoeuvres, the control strategy determines the instantaneous spark timing necessary to provide the required level of torque modulation by interrogating a brake torque response surface model stored in the engine control computer in look-up table form. Typically, these look-up tables are populated or *calibrated* using output from response surface models developed during the course of an engine mapping study.

Due to the empirical nature of the modelling, engine mapping methods rely heavily on statistical methods. In particular, Holliday [1, 2] has demonstrated the applicability of *2-stage regression* techniques for the analysis of engine mapping data. Holliday's unique 2-stage approach built on previous theory found in the literature concerned with the modelling of *repeated measurements* data [9, 10, 11]. As will be seen, the applicability of this 2-stage approach to the modelling follows directly from the fact that engine mapping response data are organised in *sweeps*. In engine mapping studies the *data sweep* replaces the *individual* in repeated measures analysis as the basic experimental unit.

Holliday's recommended analysis procedure possesses the advantages that it accounts for the inherent structure in the data imposed by the experimental data collection method. Also, through its use of *response features* [12], it has the virtue of providing a statistical modelling framework that is more readily interpreted by engineering analysts. As will be seen, areas such as model selection, diagnostics and validation require only relatively simple procedures. Only the final multivariate parameter estimation phase, based on maximum likelihood techniques, involves any complex analysis.

However, in exemplifying his method Holliday focused on the analysis of a small data set collected over a limited range of engine operation. Consequently, only very simple models were required for predicting systematic response feature behaviour. Although sufficient to illustrate the merit of the technique, the important practical issues of selecting more realistic models and their corresponding experimental design protocols were not addressed. This is the contribution made in this paper.

2 CASE STUDY DESCRIPTION

The engine-mapping model describes the behaviour of the engine as a function of the primary engine state and actuation variables. Typically, the responses of interest are the legislated exhaust emissions, exhaust temperature, standard deviation of indicated mean effective pressure, indicated torque and brake torque. However, to simplify the presentation, in this paper only the modelling of brake torque is considered. Specifically the focus will be to accurately model brake torque as a function of (N, L, A, S). The process undertaken is depicted conceptually in figure 1. Similarly, figure 2 presents the corresponding two-stage modelling process.

In this engine mapping exercise the fundamental experimental unit is the spark sweep. During the sweep (N, L, A) are held constant and the various characteristics of interest are recorded

C606/025/2002 © Ford Motor Company 2002

while spark advance is varied from its minimum to its maximum value. The data set utilised in this study consists of 49 spark sweeps collected on a Ford 1.0L 8-valve single overhead camshaft engine. Each sweep consists of roughly ten (S, T) pairs, – spark knock, combustion stability or exhaust temperature considerations permitting. Before an observation is recorded the responses are allowed to settle. Consequently, correlations between successive observations within a sweep are assumed negligible.

A "V"-optimal [13] experimental design algorithm implemented in the Mathworks' Model Based Calibration Toolbox® (MBC) was employed to generate the (N, L, A) test plan. The experimental design procedure is discussed at length in section 4. A spark sweep was collected at each (N, L, A) combination specified by the design. Use of optimal experimental design procedures ensures efficient recovery of the proposed stage-2 model, which in contrast to Holliday's work, is the hybrid B-spline-polynomial proposed by Lewis and Grove [14, 15].

During testing, the engine state is monitored by a substantial instrumentation package that provides a direct indication of the data quality as well as a *watchdog* facility to prevent engine damage. For example, catalyst temperature is monitored continuously and the sweep terminated if the measured value exceeds a configurable threshold. Similarly, combustion pressure transducers facilitate the study of aberrant combustion phenomena and knock intensity measurement [16, 17]. These auxiliary measurements are recorded along with the response variables of interest and are available to the analyst. As discussed in section 5 analysis of these auxiliary variables, in conjunction with the responses of interest on a sweep specific basis, provides the analyst with an invaluable data quality diagnostic.

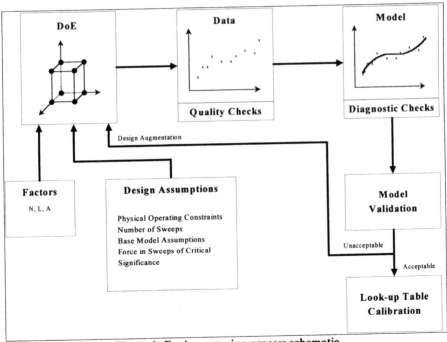

Figure 1: Engine mapping process schematic

In practice the data diagnostic checks are conducted concurrently with the stage-1 modelling, which involves fitting curves to the individual sweep response profiles. This yields 49 separate models, one for each sweep that describes the systematic behaviour in brake torque as spark advance is varied. At this juncture, the desired response features are calculated from knowledge of the stage-1 fit coefficients. Analysing changes in the response features among sweeps, *i.e.* as (N, L, A) vary, represents the primary focus of the second stage model. This process is itself subdivided into two component parts: univariate and multivariate analysis.

Univariate analysis refers to the process of selecting the respective response feature models. To simplify this each response feature is considered separately. Since the form of the full second order model is linear, straightforward stepwise regression techniques suffice. Although a variety of standard techniques are available - e.g. forward, backwards or stepwise regression methods [18] – experience has shown that the prediction error sum of squares (PRESS) [18, 19] provides an effective model selection criterion. The MBC toolbox contains a stepwise tool that implements a PRESS search. Univariate stage-2 analysis will be discussed in section 6.1. Having obtained the form of the stage-2 models using univariate methods the corresponding parameters are estimated using multivariate maximum likelihood techniques. Parameter estimation via maximum likelihood is the subject of section 6.2.

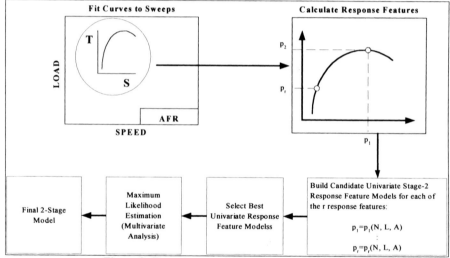

Figure 2: Two-stage modelling process schematic

Having fitted the model to the data, it is essential to conduct an assessment of its validity prior to use. Primarily this involves the collection of fresh data with which to investigate the model's predictive performance. The process of predicting fresh data is referred to as *external validation* in this study. In addition, considerable attention is paid to the model coefficients, their sign and ensuring generally that the model behaves in accordance with physical theory. In this endeavour much use is made of graphical techniques for visualising the response

C606/025/2002 © Ford Motor Company 2002

surface behaviour. Here, this is referred to as *internal validation*. Typically, internal validation proves an important step. Even in circumstances where the external validation results are satisfactory, in the author's experience, models that do not exhibit the expected physical trends are rarely acceptable to engineers.

3   TWO-STAGE MODEL DESCRIPTION

The presence of repeated observations on a data sweep requires special care in characterising the variation in the experimental data. In particular, it is important to represent two sources of variation: random variation among measurements *within* a sweep (*intra-sweep variation*) and random variation *among* sweeps (*inter-sweep variation*). Inferential procedures accommodate these different variance components within the framework of an appropriate hierarchical statistical model. Often the choice of the curve fit function at stage-1 is motivated entirely by empirical considerations. Consequently, parameters associated with the stage-1 model often have no direct engineering interpretation. Rather, it is characteristic geometric features of the curve or *response features* [12] that are of interest.

Since response features are chosen specifically to have an engineering interpretation it is preferable to model the variation in the response features across sweeps rather than the curve-fit coefficients themselves. This results in models that are better understood by engineering analysts. Furthermore, as discussed by Holliday [1, 2], the decomposition of the modelling into two stages simplifies the overall analysis. Attention is now focused on the formulation of the statistical model. Davidian and Giltinan's text provides [9] a detailed treatment of the 2-stage non-linear repeated-measures model. Here only a brief description is provided for the purposes of continuity.

Let T_{ij}, denote the jth measured brake torque value, $j \in (1, n_i)$, for the ith spark sweep, $i \in (1, m)$. Similarly let the corresponding spark advance value be denoted by S_{ij}. At stage-1 it is assumed that:

$$T_{ij} = f_i(S_{ij}, \beta_i) + e_{ij} \qquad\qquad 1$$

where $\beta_i \in \Re^r$ is a sweep-specific parameter vector and e_{ij} is a random error term reflecting uncertainty in the response with $E(e_{ij} | \beta_i) = 0$. The notation $f_i(S_{ij}, \beta_i)$ allows for the fact that the exact form of the stage-1 fit function may vary from sweep-to-sweep. The necessity for this will become apparent in section 2.2.

Collecting the responses and errors for the ith sweep into the vectors $\mathbf{T}_i = [T_{i1}, \ldots, T_{in_i}]^T$ and $\mathbf{e}_i = [e_{i1}, \ldots, e_{in_i}]^T$ respectively and defining the vector $\mathbf{f}_i(\beta_i) = [f_i(S_{i1}, \beta_i), \ldots, f_i(S_{in_i}, \beta_i)]^T$, the data for the ith sweep can be summarised in the compact form:

$$\mathbf{T}_i = \mathbf{f}_i(\beta_i) + \mathbf{e}_i, \qquad\qquad \mathbf{e}_i | \beta_i = \mathcal{N}_{n_i}[\mathbf{0}, \sigma_i^2 \mathbf{I}_{n_i}] \qquad\qquad 2$$

where $\mathcal{N}$ denotes the normal distribution. Equation 2 suggests a particularly simple intra-sweep covariance model. Although in principle the analyst is free to select more complex structures to account for heterogeneous variance experience has shown that for modelling brake torque data this is rarely necessary. Davidian and Giltinan [9, 20] discuss more general covariance models and a method of parameter estimation based on iterative generalised least squares techniques.

Once the desired stage-1 model is fitted to the sweep-specific data, attention naturally focuses on the calculation of the relevant response features. Since the response features are specifically chosen to have an engineering interpretation it is preferable to model the variation in the response features across sweeps rather than the β_i themselves. In general, the response features will be related to the fit parameters through a non-linear vector valued function, $p_i(\beta_i)$ say. Thus, the stage-2 modelling is concerned with relating the systematic variation in the $p_i(\beta_i)$ to changes in the remaining engine parameters; i.e. (N, L, A) in this case.

In this study it is assumed that at stage-2 the response features can be approximated by a form linear in (N, L, A), but not S, with additive independent errors b_i having common covariance matrix D. Thus:

$$p_i = a_i\theta + b_i, \qquad\qquad b_i \sim \mathcal{N}_r(0, D) \qquad\qquad\qquad 3$$

where the model specific matrix a_i depends on (N_i, L_i, A_i), the mean values of engine speed, load and air-fuel-ratio for the ith spark sweep. Similarly, θ is the stage-2 model parameter vector. A discussion of the precise form of $a_i\theta$ is deferred until section 2.3.1.

The primary weakness of equation 3 is its failure to account for the uncertainty encountered in estimating p_i. Incorporation of this uncertainty is usually based on the asymptotic theory for $p_i | \theta$. It is assumed that $p_i | \theta$ is approximately $\mathcal{N}(p_i, \sigma_i^2\Sigma_i)$, where $\sigma_i^2\Sigma_i$ is the asymptotic covariance matrix for p_i. Under this assumption, the marginal distribution of p_i is approximately normal with mean $a_i\theta$ and covariance matrix $W_i = \sigma_i^2\Sigma_i + D$, so that p_i can be written:

$$p_i = a_i\theta + b_i + e_i^* \qquad\qquad\qquad 4$$

where e_i^* is assumed to be $\mathcal{N}(0, \sigma_i^2\Sigma_i)$. This approximation suggest that both θ and $D(\omega)$ can be estimated by minimising the negative log likelihood function:

$$L(\theta, \omega) = \sum_{i=1}^{m} \left(\log|W_i| + (p_i - a_i\theta)^T W_i^{-1}(p_i - a_i\theta) \right) \qquad\qquad 5$$

where ω is a vector comprised of the $r(r+1)/2$ unique elements of D. In 5, for notational compactness the dependency of W_i on ω has been suppressed. When implementing this approach, an estimate of the asymptotic covariance matrix $\sigma_i^2\Sigma_i$ for the ith sweep would be used. The estimate of $\sigma_i^2\Sigma_i$ would be then treated as fixed during the subsequent minimisation of 5.

Although estimation of θ and ω appears predicated on the approximate normality of the distribution of $p_i | \theta$, the method has omnibus appeal. The significant feature is that the first two marginal moments of p_i may be written as $a_i\theta$ and $\sigma_i^2\Sigma_i + D$ respectively. This follows directly from equation 4, which is linear in b_i and e_i^*. Given this approximately linear

C606/025/2002 © Ford Motor Company 2002

marginal model, the estimator for θ has the form of a generalised least squares estimator. Consequently, equation 5 proves a natural choice as a basis for estimation of θ and ω even if the distributions of $\mathbf{b}_i$ and $\mathbf{e}_i^*$ are not exactly normal.

3.1 Physical interpretation of the variance components

Immediately apparent is that equations 2 and 3 reveal that response data are subjected to two sources of variation: variation within sweeps and variation among sweeps, represented by $\mathbf{e}_i$ and $\mathbf{b}_i$ respectively. The intra-sweep error is readily interpreted as the inherent *measurement noise* superimposed on the signal under study. Similarly, the inter-sweep error also has a physical interpretation.

When the engine is set up to undertake a spark-sweep there exist certain nuisance factors, such as humidity or inlet temperature and so on, that have a small influence on the sweep and so the corresponding estimate of the response features. Such factors vary among spark sweeps but are considered constant for the duration of a sweep. If the analyst were to repeat a sweep these nuisance variables would take on new values perturbing the sweep yet again, yielding new response feature estimates. Information regarding the aggregate influence of these nuisance variables on the response feature estimates is contained in the $\mathbf{b}_i$. Since each time a sweep is repeated the nuisance variables take on random values, the associated $\mathbf{b}_i$ can be considered to be a vector of *random effects* that measure the perturbation of the observed response feature estimates from the true population average. Thus, specifying the distribution of the $\mathbf{b}_i$ provides a convenient means of summarising the influence of *test-to-test* variation on the sweep specific response feature estimates.

3.2 Stage-1 model description

Figure 3 presents a typical spark sweep selected from the data set. The curve exhibits a well-defined maximum and asymmetric curvature either side of the stationary point. At stage-1, the primary focus of the modelling effort is to find a fit function that adequately describes the salient features of the spark sweep. This model is subsequently utilised to calculate the necessary response features that are the subject of the second-stage modelling.

As demonstrated by the plot, the asymmetric nature of the stage-1 response characteristic indicates that use of a quadratic curve fit function is inappropriate. Although quadratic curves have often been utilised for fitting spark sweep data in the author's experience they often exhibit significant bias. As an alternative Holliday has suggested using a segmented polynomial [21] or spline model. This approach possesses the merit of simplicity as well as being able to more adequately describe the important features of individual spark sweeps. Holliday's suggested stage-1 model has the following form:

$$E(T_{ij} \mid \mathbf{b}_i) = \beta_{0i} + \beta_{LQi}(S_{ij} - k_i)_-^2 + \beta_{RQi}(S_{ij} - k_i)_+^2 \qquad 6$$

where $(S_{ij} - k_i)_- = \min\{0 \ (S_{ij} - k_i)\}$ and similarly $(S_{ij} - k_i)_+ = \max\{0 \ (S_{ij} - k_i)\}$. Again experience has shown that for some sweeps at high levels of spark retard equation 6 exhibits significant bias. The ability to accurately predict brake torque under these circumstances is important for the control of driveline *clunk* and *shuffle* during throttle transients [22]. This issue has been resolved by adding a cubic term as follows:

$$E\left(T_{ij} \mid \mathbf{b}_i\right) = \beta_{0i} + \beta_{LCi}\left(S_{ij} - k_i\right)_-^3 + \beta_{LQi}\left(S_{ij} - k_i\right)_-^2 + \beta_{RQi}\left(S_{ij} - k_i\right)_+^2 \qquad 7$$

Not all the spark sweeps are of the form presented in figure 3. Several of the sweeps were terminated prior to achieving maximum brake torque, or just after, usually due to the onset of moderate or heavy detonation. Similarly, some sweeps exhibit only limited information to the left of the maximum where exhaust gas temperature or combustion stability constraints apply. In such cases equation 7 is inappropriate.

To resolve this issue, it is first necessary to appreciate that equation 7 places no restrictions on how the sweep specific response features are actually calculated. This implies that the stage-1 model can vary among sweeps, with the proviso that the desired response features can still be calculated from the coefficients associated with the alternate model.

For example, assume that one response feature of interest is the spark advance yielding maximum brake torque. Let this spark timing be denoted by MBT. Then, by inspection, equation 7 implies that $MBT_i = k_i$. Now assume that the sweep in question is limited by the onset of detonation and that maximum brake torque is in reality not achieved. Under these circumstances the MBC software fits the quadratic relationship $\beta_{0i} + \beta_{Li}S_{ij} + \beta_{Qi}S_{ij}^2$ to the data for the ith sweep. MBT_i is now calculated using the relation $MBT_i = -\beta_{Li}/2\beta_{Qi}$, which may represents substantial extrapolation of the data. In addition, if the quadratic relationship is used to summarise the sweep behaviour then the corresponding form of the covariance matrix Σ_i must also be modified. This has obvious implications for the estimation of θ and ω via minimisation of equation 5. The MBC software automatically calculates and applies the appropriate Σ_i dependant on the choice of $\mathbf{f}_i\left(\beta_i\right)$.

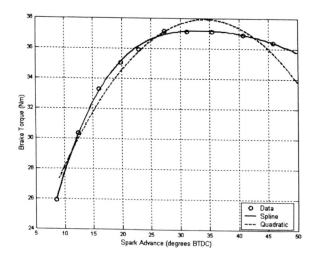

Figure 3: Example spark sweep and curve fit comparison

C606/025/2002 © Ford Motor Company 2002

3.2.1 A note on the conditioning of the stage-1 Jacobian matrix

An important problem in linear regression, usually referred to as multicollinearity [18], occurs when some of the columns of the regression matrix are highly correlated. Strong multicollinearity results in large variances and covariances for the least squares estimators of the regression coefficients. The characteristic roots or eigenvalues of the information matrix can be used to measure the extent of the multicollinearity in the data. Consequently, some analysts examine the condition number for the information matrix, defined as the ratio of the maximum to minimum eigenvalues, interpreting it as a multicollinearity diagnostic.

Similar concerns exist for non-linear regression models [23]. In the non-linear case the information matrix is $\left[\mathbf{J}_i^T(\hat{\beta}_i)\mathbf{J}_i(\hat{\beta}_i)\right]$, with $\mathbf{J}_i(\hat{\beta}_i)=\partial\mathbf{f}_i/\partial\beta_i$ evaluated at $\hat{\beta}_i$. Here the situation is more complicated because the information matrix depends upon the design as well as the estimates for $\hat{\beta}_i$. Again the condition number acts as a useful diagnostic. If the condition number is large, then the standard errors for the elements comprising $\hat{\beta}_i$ will also be large. In extreme cases the estimates will be very sensitive to the sweep-specific data and the individual estimates will tend to vary wildly across sweeps. This can have serious consequences for the second-stage modelling.

3.2.2 Response feature selection

Paradoxically, although the modelling paradigm presented affords considerable flexibility in terms of specifying differing $\mathbf{f}_i(\beta_i)$ for separate sweep profiles and the corresponding response features $\mathbf{p}_i(\beta_i)$, a single model must be postulated for predictive purpose. To understand this consider the problem of calculating predicted brake torque at a new (N_i, L_i, A_i, S_{ij}) configuration.

- Use the stage-2 model to calculate the response feature vector, $\hat{\mathbf{p}}_i(\beta_i)$, evaluated at ($N_i$, L_i, A_i).
- Invert the function $\hat{\mathbf{p}}_i(\beta_i)$ to calculate the corresponding $\hat{\beta}_i$. This is equivalent to fixing the function $\mathbf{f}_i(\beta_i)=\mathbf{f}(\beta_i)$ for all sweeps.
- Given $\hat{\beta}_i$ and S_{ij} determine $\hat{T}_{ij}=f(S_{ij},\hat{\beta}_i)$.

Hence, for prediction purposes it is necessary to assume a single form for $\mathbf{f}_i(\beta_i)$. Given calculating $\hat{\beta}_i$ involves inverting the function $\hat{\mathbf{p}}_i(\beta_i)$, it is also desirable that $\hat{\mathbf{p}}_i^{-1}(\beta_i)$ is simple to evaluate and single valued. In this study, the segmented polynomial defined by equation 7 is chosen as $f(S_{ij},\hat{\beta}_i)$. Consideration will now be given to the selection of appropriate response features.

Curve fit coefficients k_i and β_{0i} have a direct engineering interpretation and can thus be considered response features by definition; k_i is the spark advance that yields maximum brake torque (denoted by MBT$_i$) and β_{0i} the corresponding peak torque value (denoted by PKTQ$_i$). Three additional response features are utilised to describe the influence of retarding or advancing the spark timing from MBT$_i$. These are ΔLESS10$_i$, ΔLESS25$_i$ and ΔPLUS10$_i$ defined as the observed torque reduction from the maximum at spark timings 10 and 25 degrees retarded from MBT$_i$ and 10 degrees advanced from MBT$_i$ respectively.

ΔLESS10$_i$ and ΔPLUS10$_i$ are useful for providing the analyst with a measure of the asymmetry of the sweep-specific brake torque response profile about the maximum. Normally

UNIVERSITY OF HERTFORDSHIRE LRC

the inequality $|\Delta \text{LESS10}_i| > |\Delta \text{PLUS10}_i|$ would be expected to hold. In addition, equivalence of these two quantities indicates that MBT was not achieved; a quadratic fit yields symmetric curvature. Similarly, ΔLESS25_i provides the engineering analyst with a sense of how rapidly the brake torque response falls away under heavy spark retard.

Using this nomenclature, for the ith sweep the corresponding response feature vector is written:

$$\mathbf{p}_i = \begin{bmatrix} \text{MBT}_i \\ \text{PKTQ}_i \\ \Delta \text{LESS10}_i \\ \Delta \text{LESS25}_i \\ \Delta \text{PLUS10}_i \end{bmatrix} = \begin{bmatrix} k_i \\ \beta_{0i} \\ 10^2 \beta_{LQi} - 10^3 \beta_{LCi} \\ 25^2 \beta_{LQi} - 25^3 \beta_{LCi} \\ 10^2 \beta_{RQi} \end{bmatrix}$$

These relationships are simple to compute and easily inverted as required.

3.3 Inter-sweep variation

Systematic variation in the response features is summarised at stage-2 using the hybrid B-spline polynomial models suggested by Lewis [14, 15]. In keeping with Lewis' approach the single spline factor is engine speed, whereas the effects of normalised air charge and air-fuel ratio are summarised by polynomial terms up to third order. A separate hybrid B-spline model is fitted to each response feature. The knot locations are distinct; consequently, the polynomial segments comprising the spline function are joined so that the first $(d-1)$ derivatives of the spline are continuous [24, 25] – where d is the degree of the polynomial segments. In this study, it is assumed that $d = 3$ (cubic polynomial segments) and there are two knots, *i.e.* $t = 2$. Consequently, for this study there are $d + t + 1 = 6$ basis functions associated with the spline factor.

By adapting the notation of Grove and Lewis [15] (GL), for this case study, the stage-2 model for the rth response feature can be written:

$$p_r = \sum_{s=1}^{6} \alpha_{s0} B_s(x_1) + \sum_{u=2}^{3} \sum_{s=1}^{6} \alpha_{su} x_u B_s(x_1) + \sum_{u=2}^{3} \sum_{v \geq u}^{3} \beta_{uv} x_u x_v$$

$$+ \sum_{u=2}^{3} \sum_{v \geq u}^{3} \sum_{w \geq v}^{3} \beta_{uvw} x_u x_v x_w + \sum_{u} \sum_{v \geq u}^{3} \beta_{1uv} x_1 x_u x_v \qquad\qquad 8$$

where $x_1 = N_i$, $x_u, x_v, x_w = L_i$ or A_i as appropriate, $B_s(x_1)$ denotes a B-spline basis function, α is a regression coefficient involving a spline basis function and β represents a regression coefficient associated with a monomial term. Therefore, each of the r stage-2 response feature models involves 28 separate terms as follows:

- 18 terms involving basis factors, $B_s(N_i)$, or interactions between basis factors and non-spline factors, *e.g.* $B_s(N_i)L_i$.
- 3 second order monomial terms: L_i^2, A_i^2 and $L_i A_i$.
- 4 third order monomial terms not involving the spline factor: $L_i^3, A_i^3, L_i^2 A_i$ and $L_i A_i^2$.
- 3 third order terms involving the spline and non-spline factor: $N_i L_i^2, N_i A_i^2$ and $N_i L_i A_i$.

C606/025/2002 © Ford Motor Company 2002

Thus the spline factor $\left(x_1 = N_i\right)$ enters the model in two forms: as a basis factor and also as a monomial term. This is the origin of the term *hybrid B-spline polynomial*. Note there is no intercept and no linear term in the non-spline factors in 8 because the basis functions must satisfy the constraint:

$$\sum_{s=1}^{b} B_s\left(x_1\right) = 1 \qquad\qquad 9$$

where b is the number of basis functions. Similarly, summation across all b columns of $B_{SL}\left(N_i\right)L_i$ corresponding to the ith sweep yields L_i. As discussed by GL such hidden relationships between terms imply that the estimated regression coefficients do not have the usual straightforward interpretation. Consequently, GL recommend these models be used primarily for prediction.

3.3.1   Full specification for $\mathbf{a}_i \theta$

If the number and location of the knots are assumed fixed, then equation 8 can be interpreted as a linear combination of known basis functions. Hence, using vector notation 8 can be written in the form:

$$p_{r,i} = \mathbf{a}_{r,i}\theta_r \qquad\qquad 10$$

where $p_{r,i}$ symbolises the rth observed response feature for the ith sweep, $\mathbf{a}_{r,i}$ is the row vector of B-spline and monomial basis function values, *i.e.*:

$$\mathbf{a}_{r,i} = \left[B_1\left(N_i\right) \quad \ldots \quad B_6\left(N_i\right)A_i \quad L_iA_i \quad \ldots \quad N_iL_iA_i\right]$$

and θ_r is the corresponding regression coefficient vector, *i.e.*:

$$\theta_r = \left[\alpha_{10,r} \quad \ldots \quad \alpha_{6A,r} \quad \beta_{LA,r} \quad \ldots \quad \beta_{NLA,r}\right]^{\mathsf{T}}$$

For the moment the full model of equation 8 is assumed for all response features. This notation is now applied to write out the second stage model for the ith sweep in its entirety. After recalling the definition of the response features in section 3.2.2 and then writing the response feature vector $\left(\mathbf{p}_i\right)$, model specific matrix $\left(\mathbf{a}_i\right)$ for the ith sweep and stage-2 parameter vector $\left(\theta\right)$ out in full yields:

$$\begin{bmatrix} MBT_i \\ PKTQ_i \\ \Delta LESS10_i \\ \Delta LESS25_i \\ \Delta PLUS10_i \end{bmatrix} = \begin{bmatrix} \mathbf{a}_{MBT,i} & \mathbf{0} & \mathbf{0} & \mathbf{0} & \mathbf{0} \\ \mathbf{0} & \mathbf{a}_{PKTQ,i} & \mathbf{0} & \mathbf{0} & \mathbf{0} \\ \mathbf{0} & \mathbf{0} & \mathbf{a}_{\Delta LESS10,i} & \mathbf{0} & \mathbf{0} \\ \mathbf{0} & \mathbf{0} & \mathbf{0} & \mathbf{a}_{\Delta LESS25,i} & \mathbf{0} \\ \mathbf{0} & \mathbf{0} & \mathbf{0} & \mathbf{0} & \mathbf{a}_{\Delta PLUS10,i} \end{bmatrix} \begin{bmatrix} \theta_{MBT} \\ \theta_{PKTQ} \\ \theta_{\Delta LESS10} \\ \theta_{\Delta LESS25} \\ \theta_{\Delta PLUS10} \end{bmatrix} \qquad 11$$

where $\mathbf{0}$ is an appropriately dimensioned row vector of zeros.

4   OPTIMAL EXPERIMENTAL DESIGN

Having discussed the form of the second stage model, consideration is now given to the corresponding optimal experimental design procedures. Here the primary emphasis centres on practical aspects critical to the successful outcome of engine mapping studies, rather than the details and performance of the optimal search algorithm itself. However, before proceeding, note the choice of the design criterion, *i.e.* V-optimality, is entirely consistent with the underlying goal of prediction. V-optimal designs attempt to minimise the average prediction error variance over the region of operability [13]. It is assumed throughout that the full model defined by equation 8 is used in evaluating the V-optimality measure.

The necessity to deviate from standard response surface designs intended to support low order polynomial models [26, 27] stems from:

- The relatively complex behaviour of MBT with engine speed. As will be seen, low order polynomials are incapable of accurately representing these characteristics. This point will be discussed further in section 4.2.
- The necessity to provide accurate predictions over the entire region of operability. In contrast standard response surface experimental designs are concerned with mapping out only a small part of this region, perhaps in the vicinity of a local optimum, where the response will often be adequately estimated by relatively simple functions.
- One very useful practical aspect of optimal experimental design procedures is that points of great practical significance can be forced into the design. For example, predictions around idle and peak power were of particular interest to the engineering analyst in this case study. The MBC toolbox provides a facility to automatically include such points in the design. As will become apparent this is especially important if these points are located close to or on a constraint boundary.
- The practical physical operating constraints imposed on the operation of any naturally aspirated spark ignition engine ensure that the feasible operating region is irregular in nature. Incorporating these constraints into the optimal design generation algorithm is straightforward. Constraints are treated briefly in the next section.

4.1   Constraint management

Optimal experimental design procedures search a *candidate list* consisting of feasible combinations of the factor levels. In engine mapping experiments the region of interest coincides with the *region of operability* – defined as the set that are physically realisable and not seriously damaging to the engine, emissions after treatment system or test cell equipment. Consequently, it is desirable to remove these points from the candidate set before executing the optimal search algorithm. For a naturally aspirated engine, fitted with a close-coupled three-way catalytic converter, three operating constraints are usually of interest:

a) The maximum achievable load threshold depends on engine speed.
b) At any engine speed there is a minimum load threshold below which stable combustion cannot be achieved.
c) Under high engine speed and normalised air charge operating conditions there is a leanest air-fuel-ratio above which the catalytic converter temperature operating limit will be exceeded. Unusually this constraint does not apply in this study. The

C606/025/2002 © Ford Motor Company 2002

maximum exhaust temperature achieved by the 1.0L engine under test is low compared to the maximum catalyst temperature threshold.

Additional data must be collected to identify these constraint boundaries. For example, the maximum load threshold can be determined from load data collected during wide-open throttle performance curves. This is particularly convenient because the MBC toolbox permits constraints to be represented in look-up table form. Custom testing is usually required to determine the remaining constraint boundaries. No specific details of these tests are provided.

However, it should be appreciated that constraints testing is not intended to be an extensive task. Exhaustive testing to precisely define the constraint boundary would negate the speed advantage offered by DoE based techniques. Therefore, the practical objective is to eliminate the majority of infeasible conditions from the candidate set. If subsequently during testing it becomes apparent that an infeasible operating condition has been selected, then several strategies are available. For example, if the degradation in the design properties is not severe the point can simply be eliminated from the design. Alternatively, the point can be replaced with a feasible point near to the original. The new test location can be decided by the analyst or generated automatically using the MBC software.

4.2   Selection of the knot locations

The inter-sweep model is linear provided that the number and locations of the knots remain fixed. In this section the issue of knot selection is addressed. In principle, this issue is complicated by the fact that each response feature may have its own hybrid B-spline polynomial model. However, in the author's experience the shape of the MBT response surface is by far the most complex. Consequently, if the selected model is chosen to suit MBT prediction it will certainly be adequate for the remaining response features, which exhibit simpler behaviour.

The engine under test is fitted with flush mounted combustion chamber pressure transducers and consequently measurements of *mass fraction burned angle* (MFBA) are available. Empirical rules exist that relate the mass burning profile to crank angle at MBT timing [28]; here the spark timing yielding a 50% MFBA of approximately 8-10 [° ATDC] is taken at MBT. Although providing only an approximate MBT estimate, MFBA has the virtue of being very easy and quick to evaluate. Using MFBA to estimate MBT permits the number and location of the spline knots to be determined based upon the MBT survey presented in figure 4.

All data are collected at stoichiometric air-fuel-ratio and 0.2 normalised air charge. Starting from 700 [RPM], at each engine speed the procedure is to adjust the spark timing to achieve the desired MFBA setting. Feedback control on MFBA is utilised for this purpose. This process is repeated until the maximum *(red line)* speed is achieved. Speed increments of roughly 250 [RPM] are employed yielding a fairly high-resolution profile. Typically, these data require about 4 hours to collect inclusive of test cell automation configuration.

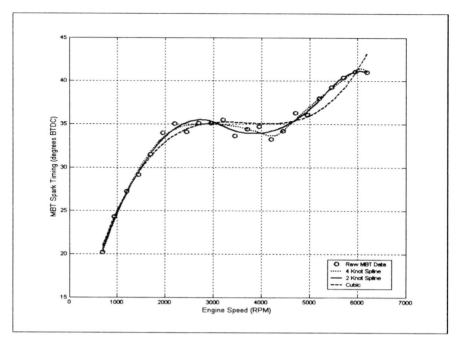

Figure 4: Selecting the number and location of the knots

To obtain the number and location of the knots a penalised free knot spline-fitting algorithm due to Lindstrom [29] is used to fit the MBT survey data. Central to Lindstrom's approach is the application of a novel multiplicative penalty function so that the penalty defines a percentage increase in the usual least squares cost function. This approach implies that the penalty is essentially scale free, allowing it to be fixed *a priori* and avoids the necessity to estimate a separate smoothing parameter for each data set.

Lindstrom's penalty function may be viewed as a minimal regularising factor that need not vary with the individual data problem. The goal is to penalise knot vectors that contain replicate and nearly replicate knots, so favouring solutions that are continuous in the derivatives. Lindstrom provides methods for selecting the penalty function and a formula for a modified generalised cross validation (GCV_{full}) criterion that accounts for the loss of degrees of freedom associated with estimating the knot positions. This provides a means of ranking prospective models.

The MBC toolbox contains a predefined model-building template for facilitating the fitting process. Several spline models may be fitted to these data and compared. Figure 4 presents the fit afforded by a two and four knot spline model and also a cubic polynomial. It is immediately apparent that the polynomial exhibits significant bias in the mid and upper speed range. In the author's estimation the 4 knot spline function over fits the data, whereas the 2-knot spline appears to fit the data well while demonstrating the necessary smoothness. The estimated knot locations are 2931 and 4165 [RPM]. These values are carried forward to the optimal experimental design procedure.

    C606/025/2002 © Ford Motor Company 2002

It is interesting to compare the GCV$_{full}$ criteria for these models; these are presented in table 1:

Table 1: Lindstrom's GCV criterion

Model	GCV$_{full}$
Cubic Polynomial	0.0599
2-Knot Spline	0.0345
4-Knot Spline	0.0323

Strictly speaking, as indicated by GCV$_{full}$, the 4-knot spline model is best. However, in the author's experience a plot of GCV$_{full}$ versus number of knots exhibits a rather flat minimum. Consequently, GCV$_{full}$ is regarded as providing a *short-list* of candidate models rather than a mechanism for automatically making a definitive choice.

4.3 Optimal design, evaluation and selection

The initial candidate list for the design was extensive being based on a 61x36x37 (N, L, A) grid. Whereas points in the L and A dimensions were equally spaced, in the speed dimension the grid density was increased in the vicinity of the knot locations [15]. The design was to be comprised of 50 spark sweeps. As the initial design is generated randomly, it is sensible to generate several candidates from which to make a final selection. Upon termination the V-optimal criterion for the selected design (V) was 0.386. Based on the author's observations made in over 50 similar studies, for useful designs $0.28 \leq V \leq 0.45$. The final 50 sweep design is tabulated in the appendix.

Of course the V-criterion is only one of several useful metrics by which the design can be judged. Consequently, incorporated into the MBC software is a comprehensive design evaluation tool that calculates many useful properties. For example, figure 5 displays the prediction error variance (PEV), scaled relative to the data, as a function of (N, L) at stoichiometric air-fuel-ratio. PEV can be interpreted as measuring the *quality* of the prediction afforded by the model – with low PEV values being associated with high quality. The authors consider PEV to be the most important metric for engine mapping designs. Note some care must be taken in evaluating these plots as the PEV may increase rapidly in extrapolated regions.

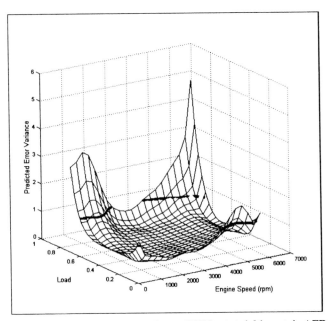

Figure 5: Prediction error variance (PEV) at stoichiometric AFR

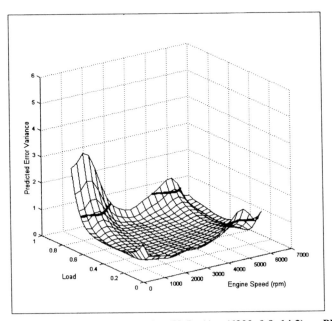

Figure 6: Effect of adding the point (N, L, A) = (6300, 0.9, 14.3) on PEV

C606/025/2002 © Ford Motor Company 2002

For regions of great engineering significance, the relevance of such plots should now be apparent. Ideally, the design should yield low PEV in these areas implying that the corresponding predictions are of adequate precision. To illustrate why it is often prudent to force in one or two points of engineering significance into the design, figure 6 illustrates the influence of adding the point (N, L, A) = (6300, 0.9, 14.3) to the design. Although artificial, as in reality this point violates both the high load and exhaust gas temperature constraints, it serves for the purposes of demonstration. As demonstrated by the figure, the effect of this change is to dramatically reduce the magnitude of the PEV in the vicinity of (6300, 0.9, 14.3).

4.4 Recovery of the Design

In engine mapping studies it is very rare for the actual sweep settings to exactly match the intended design levels. For example, depending on daily barometric pressure levels it may not be always possible to obtain intended design points that supposedly lay on the high load constraint boundary. This is because wide-open throttle load is itself a function of the observed barometric pressure. Complete knowledge of the constraint boundary is never available and it is common to find that some intended sweeps exceed combustion stability, catalyst or exhaust temperature limits. Although the optimal design algorithm can be used to select replacement points, common practice is to adjust the levels of the levels of L or A towards the centre of the design until satisfactory operation is observed. This pragmatic approach stems from the fact that, when the properties of the recovered design are compared with the intended design, rarely do such *replacement* points significantly degrade the design properties.

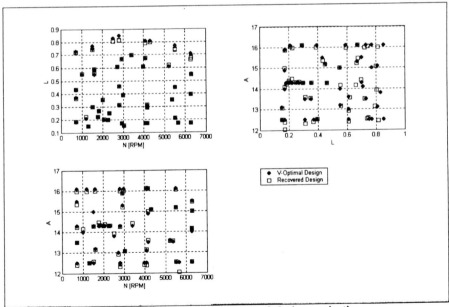

Figure 7: Original and actual stage-2 design projections

Figure 7 compares the design projections for the original 50-sweep optimal design and the 49-sweep design actually recovered. Run 50, (N, L, A) = (1500, 0.77, 15), was run at an incorrect load and air-fuel-ratio setting and was subsequently deleted from the design. This had no serious ramifications for the design properties. In the main, the actual and intended design points are in good agreement. An additional 4 sweeps were collected to serve as a small external validation data pool, making a total of 54.

5   DATA QUALITY CHECKS

To ensure the successful implementation and validity of the two-stage modelling process it is imperative to ensure that the data within sweeps is sound. For example, the analysis procedure assumes that (N, L, A) is held relatively constant during the sweep. Since the stage-2 modelling procedure is based on the mean (N, L, A) coordinates for each sweep, if one or more of these factors is permitted to drift significantly through the sweep, then the consequences for the stage-2 analysis can be grave. Thus, the first task in any sweep specific analysis is to plot the stage-2 covariate settings versus spark advance.

In addition to monitoring the stage-2 covariates it is also important to monitor the consistency of many other data channels that should remain steady throughout the experiment. For instance, measurements of load are not directly available; rather they are calculated continuously by the test-cell automation from measurements of fuel flow, air-fuel-ratio and engine speed. At low fuel consumption rates the fuel flow deviation can be unacceptably high. Consequently the reported fuel flow variation is monitored carefully throughout. In addition, since the wide band air-fuel-ratio meter employed infers this metric from exhaust oxygen concentration, it only provides a meaningful output during stable combustion.

Aberrant combustion events such as misfire or partial burn that may occur at extreme retard or advance result in artificially high exhaust oxygen concentrations. Under these circumstances, the air-fuel-meter may provide a false reading. The presence of flush mounted combustion pressure transducers and associated instrumentation means that many combustion diagnostics are available to the analyst. Consequently, the onset of abnormal combustion phenomena during the spark sweep is easily detected.

Correct component function can also be inferred from the analysis of auxiliary variables. For example, consider the three-way catalyst. During normal operation the catalyst promotes the conversion of the exhaust species into new compounds. These chemical reactions are exothermic in nature. Consequently, the catalyst mid-brick temperature should exceed the input temperature if it is working correctly. Failure of the catalyst may increase exhaust backpressure restricting the power output from the engine.

If a stage-1 spark sweep passes all data quality checks it is then modelled with the cubic spline model described in section 3.2. Auxiliary variables can also be invaluable in providing model consistency checks. For example, as discussed in section 4.2, correlations exist between 50% MFBA angle and MBT timing. Similarly, combustion stability – as determined by the standard deviation of indicated mean effective pressure usually minimizes in the

C606/025/2002 © Ford Motor Company 2002

vicinity of MBT. Engineers will expect the estimated value of k_i as defined by equation 7 to be in good agreement with these observations.

All these checks are in addition to the usual model adequacy checks based on the analysis of residuals. Plots of ordinary, weighted or studentised residuals versus predicted quantities and spark advance are also invaluable for detecting outlying data. Normal probability plots are also available. Some care must be employed in interpreting these plots when the residual degrees of freedom are small. In addition, as discussed in section 3.2 the size of the standard errors for the coefficients and associated response features are of interest. The software also calculates the condition number for the stage-1 covariance matrix.

6   STAGE-2 REGRESSION MODELLING

Focus now centres on the modelling of the response features. As outlined in section 3 the stage-2 modelling is itself divided into two stages: univariate and multivariate (maximum likelihood) analysis. The objective of the univariate analysis is to select the form of the individual response feature models. Separate multiple regression models are constructed for each response feature, which are treated for all intents and purposes as data during this process. Once the form of the response feature models is known, the response feature model parameters are determined using multivariate maximum likelihood techniques.

6.1   Univariate analysis – selection of the response feature models

Equation 8 was fitted to each estimated response feature by least squares. The MBC software automatically calculates levels for the B-spline basis functions. In doing so the software assumes the support for the basis functions corresponds to the interval $\left[\min\{N_i\}_{i=1}^m, \max\{N_i\}_{i=1}^m\right]$, where $\{N_i\}_{i=1}^m$ is the set of observed rather than intended engine speed sweep ordinates. An obvious alternative would be to utilise weighted rather than ordinary least squares. This would be consistent with the fact that the precision of the response features estimates varies among sweeps. Nonetheless, a weighted least squares approach is not pursued. Instead a PRESS minimisation stepwise regression routine embedded in the MBC software is utilised to fit each of the five univariate models. Table 2 presents the relevant summary statistics for each of the 5 regressions.

As illustrated in figure 8 MBT exhibits relatively complex trends with engine speed. The PRESS root mean square error (RMSE) is 1.782 [°], which compares favourably to an engineering requirement of 2.5 [°]. Note the similarity of the 0.2 load trend depicted in this figure compared to that in figure 4.

For PKTQ, the full model normal probability plot based on *studentised* or *t-residuals* revealed that the observed value for sweep 45 was an outlier - with a t-residual of 4.9. Consequently, this observation was set aside from the PKTQ fit. The underlying cause for this could not be discerned from the stage-1 analysis. As illustrated in figure 9, all the expected physical trends were exhibited by the PKTQ response surface. For example, dependent on (N, L) PKTQ maximises in the range $12.8 \le A \le 13.5$. Similarly, for fixed (N, A) PKTQ increases almost linearly with load.

With respect to the ΔLESS25 fit, the t-residual for observation 38 is –3. The corresponding stage-1 analysis reveals that ΔLESS25 for this sweep is an extrapolated quantity. However, setting this observation aside only marginally improves the appearance of the residual plots and does not appear to degrade the fit quality. Consequently, this observation was included in the final analysis.

For the ΔLESS10 analysis, diagnostic plots revealed observation 27 to be an obvious outlier. Again, no obvious explanation could be determined from the corresponding stage-1 sweep analysis. However, the presence of this observation appreciably degraded the fit quality and the appearance of the residual diagnostic plots. Consequently, this observation was set aside. Similar observations were made with regard to the ΔPLUS10 analysis, with this time the value from sweep 49 being removed. The ΔPLUS10 residual plots also suggest slight heteroscedastic behaviour in the variance structure, with the variance increasing with predicted ΔPLUS10. However, this is not considered a serious concern and no corrective action, such as applying a transformation to the ΔPLUS10 data, has been taken.

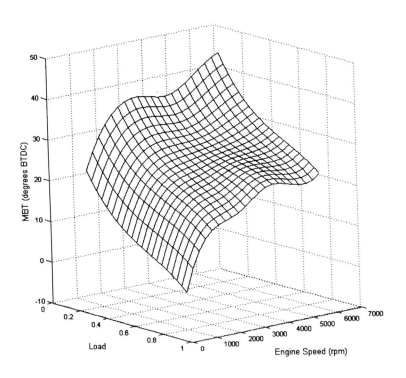

Figure 8: MBT response surface at stoichiometric air-fuel ratio

C606/025/2002 © Ford Motor Company 2002

Table 2: Univariate regression summary statistics

Response Feature	MBT	PKTQ	ALESS10	ALESS25	APLUS10
No. of Runs	49	48	48	49	48
No. of Terms	17	23	18	15	20
RMSE Residual	1.515	0.561	0.495	1.771	0.425
PRESS RMSE	1.782	0.801	0.641	2.187	0.566
R^2	0.981	0.999	0.980	0.981	0.985

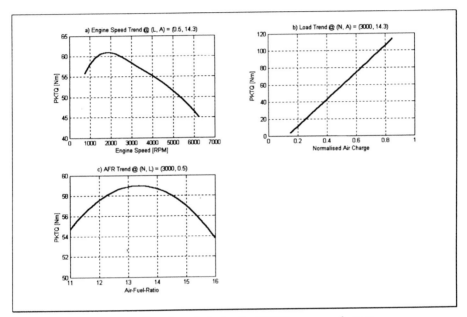

Figure 9: Trend analysis for the PKTQ response feature

6.2 Maximum likelihood (ML) estimation

An obvious weakness of the univariate analysis is that it does not account for the correlations between response features. Given the form of the individual response feature models for MBT, PKTQ, ALESS10, ALESS25 and APLUS10 determined from the univariate analysis, equation 5 can now be minimised to determine θ and ω. This process can be considerably simplified by noting that the ML estimates $\hat{\theta}$ satisfy:

$$\hat{\theta} = \left(\sum_{i=1}^{m} \mathbf{a}_i^T \mathbf{W}_i^{-1} \mathbf{a}_i \right)^{-1} \sum_{i=1}^{m} \mathbf{a}_i^T \mathbf{W}_i^{-1} \mathbf{p}_i \qquad\qquad 12$$

substitution of 12 into 5 yields the function $L(\omega)$, which can be subsequently minimised to yield the estimates $\hat{\omega}$. $\hat{\theta}$ is then calculated using 12. This approach is implemented in the MBC toolbox. As pointed out by Holliday [1], the ML estimation process is the only step in the model building process that requires the application of anything more than simple statistical techniques. Tables 3 and 4 present the associated ML estimates of the response feature variances and correlations for the study at hand.

Table 3: ML estimates of the response feature variances

Response Feature	MBT	PKTQ	ΔLESS25	ΔLESS10	ΔPLUS10
Variance	$(0.946)^2$	$(0.594)^2$	$(0.183)^2$	$(0.176)^2$	$(0.185)^2$

Table 4: Stage-2 ML estimates of the response feature correlations

	MBT	PKTQ	ΔLESS25	ΔLESS10
PKTQ	0.2396			
ΔLESS25	0.5297	-0.3899		
ΔLESS10	0.3487	0.5891	-0.1011	
ΔPLUS10	-0.1142	-0.4540	-0.1429	0.1202

Table 4 reveals there are no large correlations between the selected response features. Consequently, as illustrated in figure 10, sweep-specific brake torque estimates based on 12 do not significantly out perform those based on ordinary least squares (OLS) derived from the separate univariate analyses. Similar results were observed for the remaining sweeps utilised to train the model.

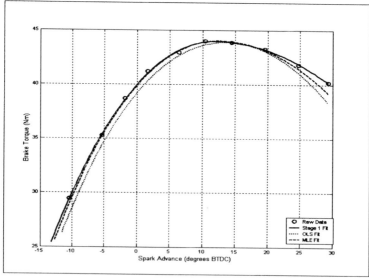

Figure 10: Comparison between the MLE and OLS or univariate fits for sweep #8

C606/025/2002 © Ford Motor Company 2002

Figure 11 presents some example fits for the MLE model to a selection of sweeps from the training data. These plots are considered representative for the entire set of 49 sweeps. Figure 12 plots the actual brake torque versus the predicted brake torque for the entire training data set. Clearly, the fit to the training data is excellent.

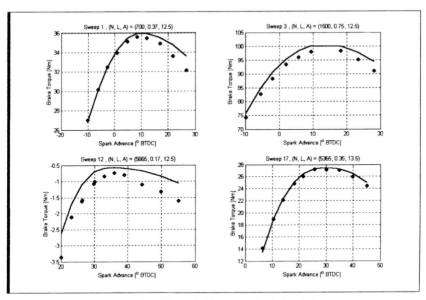

Figure 11: Example fits to the training data

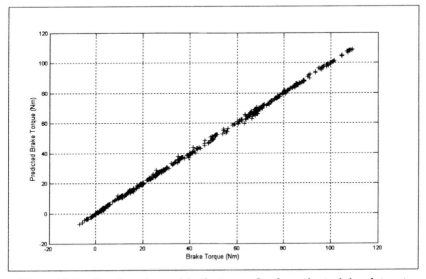

Figure 12: Predicted vs observed brake torque for the entire training data set

7   EXTERNAL VALIDATION

To gain some insight into the predictive capability of the final model, consider figure 13 that presents results from a small external validation study. For these data the root mean square error of prediction is 1.341 [Nm]. The apparent lack of smoothness in the predictions is explained by noting that each point has been calculated using the actual observed speed, load and air-fuel-ratio, rather than values of (N, L, A) averaged for the sweep. In addition, when considering these plots allowance must be made for the fact small errors in MBT or PKTQ will lead to the whole curve being offset, regardless of how well the other response features are estimated.

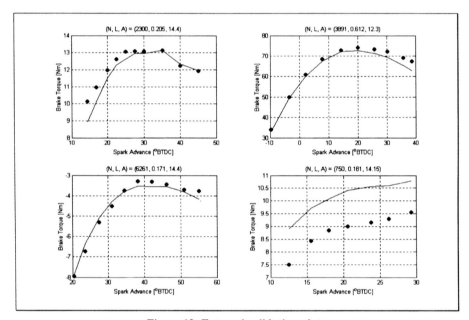

Figure 13: External validation plots

Individual predictions were calculated using the formula:

$$\hat{T}_{ij} = PKTQ_i + \left(\frac{1}{15}\right)\left[\frac{\Delta LESS10_i - PKTQ_i}{10^2} - \frac{\Delta LESS25_i - PKTQ_i}{25^2}\right]\left(S_{ij} - k_i\right)_-^3$$
$$+ \left(\left[\frac{\Delta LESS25_i - PKTQ}{25^2}\right]\left(\frac{25}{15}\right)\left[\frac{\Delta LESS10_i - PKTQ_i}{10^2} - \frac{\Delta LESS25_i - PKTQ_i}{25^2}\right]\right)\left(S_{ij} - k_i\right)_-^2$$
$$+ \left[\frac{\Delta PLUS10_i}{10^2}\right]\left(S_{ij} - k_i\right)_+^2$$

In the above expression, the dependency of MBT, PKTQ, $\Delta LESS10$, $\Delta LESS25$ and $\Delta PLUS10$ on (N, L, A) has been suppressed.

C606/025/2002 © Ford Motor Company 2002

8   DISCUSSION AND CONCLUSIONS

The model presented demonstrates the effectiveness of the 2-stage regression approach when applied to the analysis of engine mapping data, collected over the entire region of operability, for the 1.0L 8-valve engine under study. Hierarchical modelling approaches are appropriate because they directly account for the inherent structure imposed on the data by a sweep based data collection methodology. Thus, output from an engine mapping experiment should be considered a collection of spark sweeps – with the spark sweep representing the basic experimental unit.

At stage-1, the modelling is concerned with deriving a sweep-specific relationship that adequately summarises the observed systematic behaviour in brake torque with spark advance. Simple segmented polynomial or spline relationships have been shown to be adequate for this purpose. To improve the engineering interpretation of the model, the character of the sweep is summarised using response features. It is the variation in response features among sweeps that represents the focus of the second stage model.

Hybrid B-spline polynomial models have been shown to adequately describe the response feature variation among sweeps. These models prove almost ideal for this purpose; permitting the complexity of the MBT response feature with engine speed to be accurately represented, while supporting optimal experimental design procedures that permit efficient and realistic engine test plans to be developed. This permits models capable of accurately predicting the desired response features over the whole region of operability to be developed. In the authors' estimation, this represents a considerable advance over the pioneering work of Holliday. This is the main contribution made in this study.

Although in this paper attention has focused on the modelling of brake torque only, the procedure outlined is suitable for any of the important engine responses including emission and combustion metrics. Regardless of the response characteristic, the entire modelling process is easily implemented in the MBC toolbox. Indeed, such models are routinely employed at Ford for engine calibration or simulation purposes. Note, when modelling several responses it may be advantageous to employ a multi-response model that can account for the correlations between characteristics.

APPENDIX – "V"-Optimal Experimental Design

Sweep #	N	L	A
[none]	[RPM]	[none]	[none]
1	700	0.37	12.5
2	3030	0.15	13.1
3	1500	0.75	12.5
4	2800	0.31	12.5
5	2700	0.61	13
6	700	0.73	16.1
7	2800	0.85	16.1
8	700	0.43	15.5
9	700	0.73	13.5
10	4080	0.67	16.1
11	6265	0.71	14.1
12	5665	0.17	12.5
13	6265	0.39	12.5
14	4315	0.81	15.1
15	1200	0.21	16.1
16	2930	0.67	15.3
17	5365	0.35	13.5
18	6265	0.17	14
19	4165	0.29	16.1
20	4165	0.17	14.9
21	4165	0.31	13.5
22	2800	0.85	12.5
23	4080	0.81	13.1
24	2500	0.83	13.8
25	1300	0.15	12.5

Sweep #	N	L	A
[none]	[RPM]	[none]	[none]
26	5515	0.77	16.1
27	3980	0.61	12.5
28	1600	0.55	13.2
29	5515	0.77	12.5
30	5515	0.21	16.1
31	1600	0.59	16.1
32	2980	0.39	16.1
33	5215	0.61	13.6
34	2930	0.17	15.9
35	5515	0.45	15.2
36	1500	0.3	14.3
37	700	0.18	14.3
38	2050	0.2	14.3
39	2250	0.2	14.3
40	2300	0.25	14.3
41	1750	0.22	14.3
42	2000	0.35	14.3
43	1800	0.27	14.3
44	3400	0.7	14.3
45	1000	0.55	14
46	4080	0.17	12.5
47	2800	0.46	14.3
48	6300	0.7	15.5
49	6300	0.55	15
50	1500	0.77	15

C606/025/2002 © Ford Motor Company 2002

REFERENCES

1   HOLLIDAY, T., *The Design and Analysis of Engine Mapping Experiments: A Two-Stage Approach*, Ph.D. Thesis, University of Birmingham, 1995.

2   HOLLIDAY, T., LAWRANCE, A. J., DAVIS, T. P., Engine-Mapping Experiments: A Two-Stage Regression Approach, *Technometrics*, 1998, Vol. 40, pp 120-126.

3   HOLLIDAY, T., LAWRANCE, A.J. AND DAVIS, T.P. *A non-linear two-stage model approach to engine mapping*; Proceedings of the 11th International Workshop on Statistical Modelling, Orvieto, Italy. (1996), 208-215.

4   DAVIS, T.P. AND LAWRANCE, A.J. *Engine mapping: a two stage regression approach based on spark sweeps*. In Statistics for Engine Optimization (2000). S.P.Edwards, D.M. Grove & H.P. Wynn (eds), 99-108. Professional Engineering Publishing (IMechE).

5   HENDRICKS, E., SORENSON, S. C., *Mean Value Modelling of Spark Ignition Engines*, SAE Technical Paper 900616, 1990.

6   CROSSLEY, P. R., COOK, J. A., *A Nonlinear Engine Model for Drivetrain System Development*, International Conference on Control '91, Conference Code14700, Edinburgh UK, 25-28 March 1991, IEE Publication 332, Vol 2, pp. 921-925.

7   KRENZ, R. A., *Vehicle Response to Throttle Tip-In/Tip-Out*, SAE Technical Paper 850967, 1985.

8   JANSZ, N. M., DELASALLE, S. A., JANSZ, M. A., WILLEY, J., LIGHT, D. A., Development of Driveability for the Ford Focus; a Systems Approach Using CAE, *1999 FISITA EAEC Congress*, Barcelona, June 30 - July 2 1999, Paper number STA99C214

9   DAVIDIAN, M., GILTINAN, D. M., *Nonlinear Models for Repeated Measurement Data*, Chapman & Hall, First Edition, 1995.

10   DAVIDIAN, M., GILTINAN, D. M., Analysis of repeated measurement data using the nonlinear mixed effects model, *Chemometrics and Intelligent Laboratory Systems*, 1993, Vol. 20, pp 1-24.

11   DAVIDIAN, M., GILTINAN, D. M., Analysis of repeated measurement data using the nonlinear mixed effects model, *Journal of Biopharmaceutical Statistics*, 1993, Vol. 3, part 1, pp 23-55.

12   HAND, D. J., CROWDER, M. J., *Practical Longitudinal Data Analysis*, Chapman and Hall, First Edition, 1996.

13   ATKINSON, A. C., DONEV, A. N., *Optimal Experimental Designs*, Oxford: Oxford Science Publications, First Edition, 1992.

14   LEWIS, S. M., *Factorial Designs for Engine Mapping Experiments*, University of Southampton Technical Report 319, Feb. 1999.

15   GROVE, D. M., LEWIS, S. M., *Analysis of Engine Mapping Data Using B-Splines*, IMechE Conference on Statistics and Analytical Methods in Automotive Engineering, 2002, London UK.

16   BRUNT, M. F. J., EMTAGE, A. L., *Evaluation of IMEP Routines and Analysis Errors*, SAE Technical Paper 960609, Feb 1996

17   BRUNT, M. F. J., POND, C. R., BIUNDO, J., *Gasoline Engine Knock Analysis using Cylinder Pressure Data*, SAE Technical Paper 980896, Feb 1998

18   MONTGOMERY, D. C., PECK, E. A., *Introduction to Linear Regression Analysis*, 1992, Wiley Inter Science, Second Edition.

19   ALLEN, D. M., *The Prediction Sum of Squares as a Criterion for Selecting Variables*, University of Kentucky Technical Report No. 23, 1971.

20 DAVIDIAN, M., GILTINAN, D. M., Some Simple Methods for Estimating Intraindividual Variability in Nonlinear Mixed Effects Models, *Biometrics*, 1993, Vol. 49, pp 59-73.

21 GALLANT, A. R., FULLER, W. A., Fitting Segmented Polynomial Regression Models Whose Join Points Have to be Estimated, *Journal of the American Statistical Society*, 1973, Vol. 68, pp 144-147.

22 JANSZ, N. M., DELASALLE, S. A., JANSZ, M. A., WILLEY, J., LIGHT, D. A., *Development of Driveability for the Ford Focus; a Systems Approach Using CAE*, 1999 FISITA EAEC Congress, Barcelona, June 30 - July 2 1999, Paper STA99C214.

23 SEBER, G. A. F., WILD, C. J., *Nonlinear Regression*, John Wiley and Sons, First Edition, 1989.

24 LINDSTROM, M. J., Penalized Estimation of Free-Knot Splines, *Journal of Computational and Graphical Statistics*, 1999, Vol. 8, part 2, pp 333-352.

25 EUBANK, R. L., Approximate Regression Models and Splines, *Communications in Statistics - Theoretical Methods*, 1984, Vol. 13, No. 4, pp 433-484.

26 BARKER, T. D., Engine Mapping Techniques, *International Journal of Vehicle Design*, 1982, Vol. 3, pp. 142-152.

27 GROVE, D. M., DAVIS, T. P., *Engineering Quality and Experimental Design*, 1992, Longman

28 HEYWOOD, J. B., *Internal Combustion Engine Fundamentals*, McGraw-Hill, First Edition, 1988.

29 LINDSTROM, M. J., Penalized Estimation of Free-Knot Splines, *Journal of Computational and Graphical Statistics*, 1999, Vol. 8, part 2, pp 333-352.

C606/025/2002 © Ford Motor Company 2002

C606/003/2002

Application of a Pareto-based evolutionary algorithm to fuel injection optimization

O ROUDENKO
Centre de Mathématiques Appliquées, Ecole Polytechnique, Palaiseau, France
C RANDAZZO and L DORIA
Department of Statistical Methodologies, Fiat Research Centre, Orbassano, Italy
E CASTAGNA
EBU–Engineering Statistics, IVECO, Torino, Italy

ABSTRACT

The paper deals with a problem of Common Rail fuel injection process optimisation. The goal of ensuring a *flexible* controlling injection process parameters (i.e. independently of engine speed and load) implies considerable increasing of the search space dimension. The approach proposed in this work consists in an independent solving a set of optimisation problems, each corresponding to some fixed ratio *speed / load*. Such consideration of the optimisation problem at hand has a number of obvious advantages, however, the question concerning the choice of the constraint values for each "local" problem makes subject of the further research.

INTRODUCTION

One of the major challenges for engine manufactures is to cope with conflicting requirements to be achieved: fuel economy and performance complying with the more stringent emission levels proposed world wide. In particular, the Euro Emission Standards have been introduced on order to progressively reduce the amount of harmful pollutants as well as vehicle noise emissions. This ever more stringent automotive based legislation represents a very important issue for engine development. In addition, the competitive scenario of the market forces industries to reduce engine development times and costs, and better fit customers needs.

Industries have been responding to these demands with the introduction of more sophisticated technical solutions. A relevant example is combustion electronic control systems in which the innovative Common Rail fuel injection equipment gives the flexibility of controlling fuel injection timing, pressure and quantity independently of engine speed and load. On the other hand, this great flexibility increases the number of design variables and hence the complexity of the optimisation process.

The paper presents some preliminary results of the Common Rail fuel injection process optimisation for heavy-duty diesel engines. In the first section the *electronically controlled engine* and the *Common Rail System* are briefly described. In the section 2, the parameterisation of fuel injection process as well as the optimisation criteria are introduced. Third section presents the evolutionary optimisation method used to solve the problem at hand. In the forth section two possible ways of the optimisation problem statement are discussed and the numerical results are presented for both of them.

1 FUEL INJECTION SYSTEM

1.1 Electronic engine

Modern engine is undergoing an important change, passing from a fully mechanical engine to an electronically controlled engine. The core of the system remains still a direct injected diesel engine with all its components. All the changes brought in by the electronic control system are referring to the engine's periphery. Such system enables the engine mechanics to much more precise behaviour and diagnostic work than any mechanical production engine.

1.2 Common Rail System

In addition, the introduction of a new technological mechanical component called *Common Rail (CR) System* allows a considerable improving of consumption and engine performances.

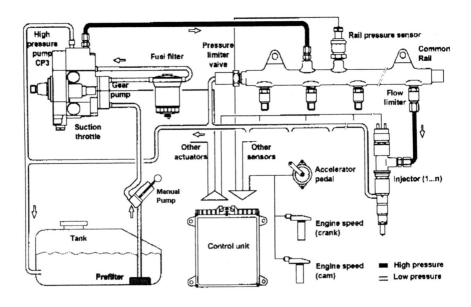

Figure 1: Common Rail System

The fuel metering of the common rail system is done by an electronically controlled lift of the injector needle. The opening time is dependent on the actually measured rail pressure. The rail pressure itself can be again electronically manipulated by the CR-pump independently of the engine speed. The set point of this rail pressure is basically defined in the ECU's data set, dependent on load and engine speed. Ambient conditions as temperatures and pressures can also have a controlled influence on the rail pressure. Therefore the system is indifferent to both, low engine speed and high fuel temperature as long as the pump capacity is high enough – the pump compensates automatically for higher losses since it controls the rail pressure. There is no bleeding necessary because the injector needle is electrically operated. The manual pump in the low-pressure circuit serves for a quicker refilling because of the general weakness of gear pumps. While the classic injection pumps have a volumetric fuel metering, the fuel metering of a CR-system is defined by the opening time of the injector needle and the rail pressure. Figure 1 shows a schematic realisation of an electronically controlled Common Rail Engine System.

1.3 Pilot injection
Generally, all electrically operated injectors have the capacity to do more than one injection per cycle. In IVECO two of them are applied on many engine models; the main injection and the pilot injection. The pilot injection means injecting a small defined quantity at a specified angle before TDC. The quantity and timing are set by the *electronic control unit* (ECU). The pilot injection helps mainly to reduce combustion noise.

1.4 Timing and ACEA Euro3 European Normative
Classical mechanical timing system is able to adapt the injection timing to the engine speed. Further there exist the devices for full-load and low idle conditions. But it is rather difficult to realize an ACEA-cycle satisfying Euro3 emission constraints with such systems. In an electronic engine control system, such parameters as main and pilot injection timing, rail pressure, injector characteristics are stored in 3D-maps. It allows an individual controlling combustion in every point of the *engine operation map* (see Figure 3) in order to satisfy the environmental normative.

2 OPTIMISATION PROBLEM

2.1 Control Parameters And Responses
Generally considered control parameters and system responses are represented in the Figure 2. In this study, we are interested in optimising following electronic injector control parameters: *timing* and *pressure* of the main injection and *timing* and *duration* of the pilot injection. The optimisation goal is to minimise specific consumption (BSFC) and combustion noise with respect to the EU Emission Standard.

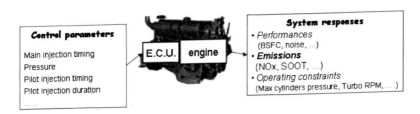

Figure 2

C606/003/2002 © IMechE 2002

As shown in the previous section, Common Rail system allows managing different behavior in combustion chamber varying control parameters in order to reduce specific consumption, noise and pollutant emissions. This flexibility of the fuel injection process control reduced into optimisation terms means that the set of four control parameters has to be optimized for each *cycle point* on the *engine map* (see an example of such map in the Figure 3). In another words, the search (decision) space dimension of an optimisation problem at hand is equal to the number of control parameters multiplied by cycle points number which may vary in function of the engine type.

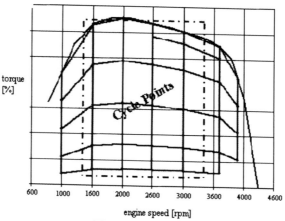

Figure 3: engine map

Since electronic injector control parameters are correlated, we deal here with a *multivariate parameter system*. In addition, the trend of combustion noise and consumption being opposite, we will search for a set of equivalent compromise solutions as any unique optimal solution exists for such kind of problems. Hence, it is now impossible to investigate combustion with *one factor at time* method.

An efficient process of solving such problems falls into three successive steps: *experimental design, statistical modeling* and *optimisation*. Respective research items point out reducing experimental effort and getting robust models, getting better approximations of complex responses and elaborating effective multi-objective optimisation techniques. In this paper, we do not present any details concerning two first stages but concentrate on the optimisation issues.

2.2 Problem Statement

Let us note x_k vector of four variables representing the fuel injection control parameters enumerated here above corresponding to the point k on the engine map. Here and there, we will use following notations :

$f_{jk}(x_k)$ – consumption ($j = 1$) or combustion noise ($j = 2$) in the control point k ;

$g_{jk}(x_k)$ – emission of the pollutant j in the cycle point k ;

c_j – Euro Emission Standard constraint value for the pollutant j;

Legislatively, the emissions are measured in each cycle point. Different points don't have the same importance: a weight associated to each of them corresponds to the percentage of time that an engine statistically works keeping a ratio speed/torque represented by a point at hand. We note

ω_k – weight of the point k.

The EU Emission Standard establishes two types of constraints. For some pollutants the emission must not exceed the limit c_j in <u>each</u> control point. That means, the following inequalities have to be verified:

$$g_{jk}(x_k) < c_j \quad \text{for each } k=1,...,K \qquad \text{(I)}$$

Here K is a cycle points number. For other pollutant types the constraints are imposed as follows:

$$g_j(x) = \sum_{k=1}^{K} \omega_k \, g_{jk}(x_k) < c_j, \qquad \text{(II)}$$

where $x = (x_1,...,x_K)$.

The optimisation problem consists in finding $x^* = (x_1^*,..., x_K^*)$ such that

$$f_{jk}(x_k^*) = \min_{x_k} f_{jk}(x_k) \qquad j=1,2; \quad k=1,...,K$$

and satisfying constraints (I) and (II). In practice, a set of operating constraints must be also taken in the consideration but in this preliminary study we do not care about them.

3 EVOLUTIONARY ALGORITHMS FOR MULTI-OBJECTIVE CONSTRAINED OPTIMISATION

3.1 Multi-Objective Evolutionary Algorithms

A number of MOEAs have been suggested in the last decade (1), (2). The primary reason for this increased interest is the ability of these algorithms to find multiple optimal compromise solutions in one single run.

Multi-objective problems aim at simultaneously optimising different objective functions. When those objectives are contradictory, there does not exist a single solution, and compromises have to be made.

A solution x is said to *Pareto-dominate* another solution y if x is not worse than y with respect to all criteria, and is strictly better than y with respect to at least one criterion. The set of all points of the search space that are not Pareto-dominated is called the *Pareto-set* of the multi-objective problem at hand: it represents the best possible compromises with respect to the contradictory objectives.

Solving the multi-objective problem amounts to choose one solution among those non-dominated solutions, and some decision arguments have to be given. Whereas any optimisation method can be used to get one single solution from the Pareto front by aggregating the different objectives with some predefined multiplicative factors, MOEAs are to-date the only algorithms that actually give a sample of the Pareto front, allowing decision makers to choose one of them with more complete information.

MOEAs are very similar to standard Evolution Algorithms, and the basic generation loop is the same (1): *select* some parents for reproduction; apply some *variation operators* (e.g. crossover, mutation) to those selected parents to generate offspring; *replace* some parents with some offspring. The difference lies in the Darwinian-like phases (selection and replacement) where, in most MOEAs, the notion of Pareto-dominance replaces the usual order relationship "has better fitness than".

However, in order to obtain the best possible sampling of the Pareto front, MOEAs also use two techniques borrowed to standard evolutionary computation: Pareto elitism and diversity preserving techniques.

Elitism consists in preserving in the populations some of the best individuals encountered so far along evolution. Of course, within the frameworks of MOEAs, ``best individuals" is replaced by "non-dominated individuals". Recent results (3) clearly show that elitism can speed up the performance of the MOEAs significantly, also it helps to prevent the loss of good solutions ones they have been found.

A crucial problem for multi-objective evolutionary approaches is to preserve diversity of solutions. Evolutionary Computation has worked out many niche techniques for that purpose but the majority of them require a user-defined parameter: this is true for classical niche techniques like sharing, with a sharing radius (4) or for more recent multi-objective specific techniques like squeezing, with the "squeeze factor"' (5). The NSGA-II (6) approach has been chosen mainly because it is free of such parameter.

3.2 Constraint Handling
Various constraint handling techniques have been proposed for Evolutionary Computation (7). For example, penalisation method; it is a very popular approach, but again it requires the a priory definition of the penalty parameters. Hence, it was decided to use the Infeasibility Objective method (8), that combines the constraint violations to give a single measure of an individuals infeasibility which is then treated as an objective in the Pareto ranking.

The resulting problem is thus a 3-objective optimisation problem: minimise all of fuel consumption, combustion noise and *infeasibility function*. In order to direct the optimisation towards the feasible solutions, it is necessary to use the *goal attainment method* which gives the priority to every feasible individual over the best infeasible one.

4 RESULTS

Numerical results presented in this section have been obtained by applying the NSGA-II algorithm to the problem of minimizing specific consumption and combustion noise under two (over five) EU Emission Standard constraints for heavy duty engines, notably NOx and

C606/003/2002 © IMechE 2002

Soot. Each of NOx and Soot responses is measured like a weighted sum over all control points (see (II), section 2.2).

The fuel injection model considered here does not envisage the Post injection stage. That means that for every control point we have only 4 optimisation parameters. The number of control points is 12.

Pareto solutions plotted in the Figures 4, 5 correspond to the scaled objective function values. Accepted approximation error of the statistical models for both objective functions is 0.01 in the present scale.

4.1 "Global" Problem
Numerical results presented in this subsection correspond to the following optimisation problem:

find $x^*=(x_1^*,...,x_K^*)$ such that

$$f_j(x^*) = \min f_j(x) = \min \sum_{k=1}^{12} \omega_k \, f_{jk}(x_k) \,, \qquad j=1,2$$

and satisfying constraints

$$g_j(x^*) = \min \sum_{k=1}^{12} \omega_k \, g_{jk}(x_k) < c_j \,, \qquad j=1,2$$

where g_1 and g_2 represent respectively NOx and Soot responses.

Here and there, this problem is referred as *global problem* because each objective function is represented by a weighted sum over all control points. Doing so reduces the objective space dimension from 24 (for the problem from the section 2.2) to 2. The possibility of plotting a Pareto front worked out by the NSGA-II considerably simplifies further results interpretation. In addition, such reformulation of the initial problem is quite natural taking in account the fact that every weight is sad to represent the "importance" of corresponding control point.

However, a big disadvantage of such problem statement is that the search space dimension is quite high (48, for the model with 12 control points and without post injection). In another words, the individuals the NSGA-II works with are the vectors from $\mathbf{R}^{48}$.

In the Figure 4 the optimisation results by NSGA-II for 11 runs are shown.

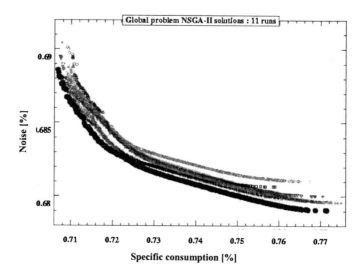

Figure 4

We can notice quite important difference between Pareto fronts obtained for different initial populations. Such instability may signify that growing up population size and generations number could improve optimisation results. But solutions presented here above have been obtained setting population size equal to 1000 and generations number to 2000 and each run took some hours on the Pentium III, 800MHz.

This is not too much time when solving a design problem but considered that this optimisation problem corresponds to a simplified model that does not take in account many factors influencing engineer decision making like operational constraints for example. In practice one may need to solve many such problems per day changing the data and studying different statistical approximation models. Evolutionary Algorithms being stochastic methods, many runs must be performed before analysing results. Hence, proposed approach is excessively time consuming.

Nevertheless, the non-dominated solutions over all runs are taken as a reference and evoked, here and there like *global problem solutions*.

4.2 Point by Point Consideration

Let us notice, that the functions f_{jk}, $k=1,...,12$ don't depend on any common variable. It suggests a very attractive perspective to decrease considerably search space dimension by solving 12 "local" problems corresponding to different control points independently. In this way, two questions have to be resolved.

First, how to represent results to make them comparable to global problem solutions? In this study, the optimisation performance of the *local problems solutions* is understood as minimising $f_j(x)$. That means that we have to assemble local solutions (vectors from $\mathbf{R}^4$) to obtain $\mathbf{R}^{48}$-vectors. In order to make such assembling process manageable, a small number of Pareto solutions of each local problem must be chosen. Combining all chosen solutions

C606/003/2002 © IMechE 2002

corresponding to different control points with each other allows us plotting the results in the plane $f_1(x) \times f_2(x)$.

The second question is how to ensure satisfying constraints of type (II) (see section 2.2)? Let us say from the beginning, that in this paper we will present rather some observations without proposing any definitive answer.

One naive solution is to fix the same constraint values for all local problems, i.e.

$$g_{jk}(x_k) < c_j \ , \qquad k = 1,...,12; \ j=1,2,$$

where c_j are the legislative constraint values for NOx and Soot. Because the sum of control point weights is equal to 1, it is a sufficient but not necessary condition for satisfying constraints (II). That means that, generally speaking, solutions of local problems with such constraint values might have worse optimisation performance than global problem solutions.

The results of this approach are not presented here not only because they are actually worse then the global problem solutions but especially because for some control points there was even no feasible solutions found. In return, we could observe that Pareto solutions worked out after 400 generations with the population size set to 200 can not be improved by further increasing algorithm parameters and that the Pareto fronts obtained for 21 different initial populations may be said to coincide. With such parameter settings, an algorithm run for every control point needs 2000 times less time respective to a global problem run. Thus, NSGA-II applied to the local problems converges (that is not the case when solving the global problem) and it runs <u>much</u> more quickly.

In fact, any choice of the constraints values for local problems potentially restricts the optimisation process respective to the global approach. However, a trial to find appropriated constraint values for each local problem has been undertaken.

The results presented below correspond to the following principle of the constraint choice. Let x be a Pareto solution of the global problem. First, $g_{jk}(x)$, $k = 1,...,12$, are calculated for all solutions x. Let us note $\tilde{g}_{jk}$ mean value of the $g_{jk}(x)$ over all x. The constraint value c_{jk} for each local problem k is chosen is chosen proportionally to the impact $\tilde{g}_{jk}$ in the global mean value $\tilde{g}_j$ and so that constraints $g_j(x) < c_j$ are satisfied. Such choice has been done assuming that the ratios $\tilde{g}_{jk}/\tilde{g}_j$ worked out during the global optimisation process ensure the best optimisation performance with respect to the Euro Emission Standard constraints.

Table 1 represents the optimisation results obtained by local and global approaches and projected in every control point subspace. The figures are ordered like correspondent control points on the engine map. The crosses represent the values $f_{jk}(x)$ for global problem solutions x. The rounds are Pareto solutions by NSGA-II applied to the local problems.

Table 1

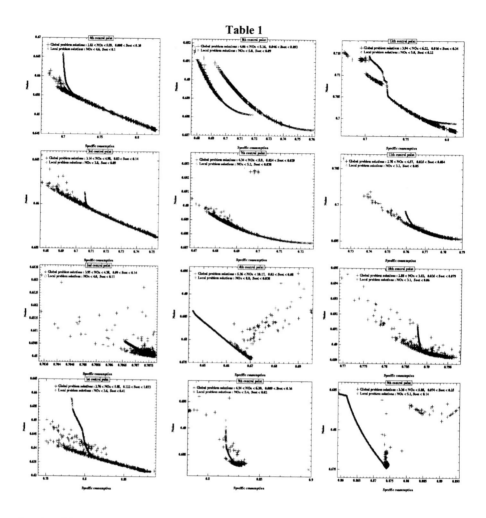

Local problems results compared to the global problem solutions are presented in the Figure 5.

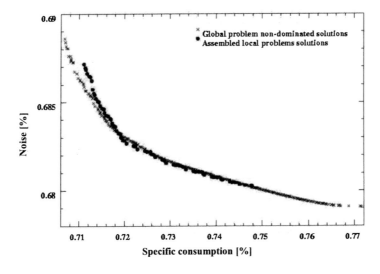

Figure 5

Let us remark that in spite of more restrictive constraints in the point by point consideration, the solutions by global and local approaches have the same optimisation performance.

4.3 Discussion

The question concerning the representation of local problems solutions is quite important. Plotting the Pareto fronts on the plane $f_1(x) \times f_2(x)$ is a convenient way to represent the results for evaluating their quality regarding all engine map.

On another hand, point by point representation is interesting while studding the design problem in more details or evaluating statistical methods used for modelling responses in different control points. When solving a real design problem, a lot of additional information must be taken in account. Hence, it may be more convenient to deal with local problem solutions and to combine them with respect to some problem issues that had not been taken in account at all in the simplified model we worked with.

The principle of the local constraints choice proposed in the previous section asks solving global optimisation problem first. Hence, it cannot be used to accelerate optimisation algorithm. The local optimisation results prove only that point by point consideration is able to provide solutions of the same performance that global approach but much faster. The main question that remains is how to find the local constraint values ensuring the best optimisation performance.

5 FUTURE WORK

The main question that remains without any satisfying answer is how to establish the local constraint values ensuring the best optimisation performance. Hence, the next step we foresee is to try to work out an algorithm based on the co-evolution principle that would allow evolving constraint values for each local problem.

On another hand, it has been observed that the rise of the NSGA-II populations towards the Pareto set of the global problem slows down quite quickly from one generation to another. We envisage using a local search procedure in order to stimulate NSGA-II evolution process, that, hopefully, will allow obtaining solutions of the same quality after much less generations.

REFERENCES

(1) Deb, K. (2001). *Multi-Objective Optimisation Using Evolutionary Algorithms*. John Wiley.

(2) Zitzler, E., Deb, K., Thiele, L., Coello Coello, C.A. and Corne, D. editors (2000). *Evolutionary Multi-Criterion Optimisation (EMO 2001)*, Springer Verlag, LNCS series Vol.1993

(3) Zitzler, E., Deb, K., Thiele, L. (2000) Comparison of Multi-Objective Evolutionary Algorithms: Empirical Results. *Evolutionary Computation*, 8(2): 173-195.

(4) Goldberg, D.E. and Richardson, J. (1987) Genetic Algorithms with sharing for miltimodal function optimisation. In Grefenstette, J.J. (Eds), *Proceedings of 2nd ICGA*, pages 41-49, Lawrence Erlbaum Associates.

(5) Corne, D.W., Knowles, J.D. and Oates, M.J. (2000) The Pareto Envelope-Based Selection Algorithm for Multiobjective Optimisation. In Schoenauer et al. (Eds.) *Parallel Problem Solving from Nature (PPSN VI)*, pages 839-848. Springer Verlag LNCS Vol.1917.

(6) Deb, K., Agrawal, S., Pratap, A., and Mayarivan, T. (2000). A Fast Elitist Non-dominated Sorting Genetic Algorithm for Multiobjective Optimisation: NSGA-II. In Schoenauer et al. (Eds.) *Parallel Problem Solving from Nature (PPSN VI)*, pages 849-858. Springer Verlag LNCS Vol.1917.

(7) Michalewicz, Z. and Shoenauer, M. (1996) Evolutionary Algorithms for Constraint Parameter Optimisation Problems. *Evolutionary Comutation*, 4(1):1-32.

(8) Wright, J. and Loosemore, H. (2001) An Infeasibility Objective for Use in Constrained Pareto Optimisation. In (2).

C606/001/2002

A parametric inlet port design tool for multi-valve diesel engines

M C BATES
Ricardo Consulting Engineers Limited, Shoreham-by-Sea, UK
M R HEIKAL
School of Engineering, University of Brighton, UK

ABSTRACT

A fully-parametric, multi-valve inlet port design tool has been developed using statistical design-of-experiments (DoE) techniques. A modular approach was adopted so that a complete port layout could be constructed using flexible, interchangeable, generic models. Additional cylinder head features and external packaging requirements were considered throughout, as were production requirements. Reliable experimental methods were used to quantify the influence of key design parameters. A predictive, knowledge-based model was developed from the DoE results and was used to predict optimum configurations for a range of scenarios. The models were successfully validated by comparing predicted results with new test data.

An overview of the tasks undertaken in developing the model is presented, including a discussion of the major findings and conclusions. An investigation using the knowledge-based model is also presented to demonstrate the practical use of the system.

1   INTRODUCTION

Inlet port flow characteristics are a critical factor in the performance of diesel combustion systems. In-cylinder flows created during the intake stroke influence bulk charge motion, turbulence generation and therefore fuel-air mixing and combustion rates. The flow capacity of the inlet ports and valves is also a key factor in determining volumetric efficiency. Increasingly stringent emissions legislation, higher customer expectations and pressure to reduce engineering costs has driven the development of new approaches to engine design. In particular, the fundamental advantages of multi-valve technology, coupled with rapidly improving fuel delivery systems has placed new requirements on inlet port performance characteristics, including low-swirl systems and small capacity hybrid engines for maximised fuel economy.

The influence of inlet ports on cylinder head design has been investigated by several authors during design studies (1,2). Inlet valve positions and port layout are major aspects of the overall design scheme and should therefore be considered at the concept stage of the design process. During this phase, care must be taken in order to meet the conflicting requirements of the various functions of the cylinder head. The development of a successful inlet port configuration remains an iterative process and the final design is often a compromise between performance and packaging. Numerous researchers have sought to identify key design parameters and their relationship with performance characteristics. The fundamental differences between directed (tangential) and helical ports have been widely studied in order to select the most appropriate type for different engine applications (3,4,5). In more recent studies, a limited number of key design features have been used to describe realistic geometry parametrically (6) and this development has been exploited to a certain extent by researchers who have used statistical methods to develop simple predictive models (7,8,9). Parametric modelling, statistical techniques and knowledge-based design are therefore emerging as potentially powerful design tools in this field of research, supported by rapid developments in computing power.

The present study was instigated to develop a predictive parametric model for multi-valve diesel engine inlet ports. The activities undertaken in meeting this objective are discussed in the next section. The most important aspects of the development process are briefly described and the experimental methods and results are presented in more detail. This paper builds on the findings of a study previously published by the authors (10) by discussing a new approach to the construction of the predictive model. Validation test results are also presented and significant trends in the model are investigated.

2   MODEL DEVELOPMENT

2.1   Parameter Selection and Definition

2.1.1 Initial Selection
Two basic port types were defined using the generic terms "directed" and "helical" to describe the overall port geometry. Because of the different swirl generation mechanisms present in each type, it was considered likely that the governing geometry parameters would be different. Therefore, it was deemed more appropriate to use two different port types in preference to a single, highly complex model. Port design parameters were selected by referring to the literature and by conducting a "brainstorming" session to develop new ideas. Features were then ranked in order of importance using a paired-comparison approach. In some cases, complex design features were investigated further in order to simplify them whilst maintaining their overall function and likely effect on flow characteristics. This was necessary to avoid the need for large numbers of parameters to describe a single feature.

2.1.2 Final design scheme – directed port
The design scheme for the directed port consisted of six parameters. Two parameters, E and At, describe the location of the inlet valve in the cylinder bore and the approach angle of the port relative to the cylinder respectively, as shown in the plan view in Figure 1(a). At may also be interpreted as the angular location of the inlet valve if the port orientation is fixed. E is a dimensionless parameter indicating the distance of the valve centre from the cylinder centre. A third parameter, Dn, was defined as the inlet valve inner seat diameter, divided by the bore diameter. The remaining parameters, Av, Ar and R, were used to describe the vertical approach angle of the port, the angle through which the port curved from entry face to inlet valve, and the radius of the curve respectively.

2.1.3 Final design scheme – helical port
The design scheme for the helical port consisted of seven parameters, four of which were common with the directed port (E, At, R and Ar). The remaining parameters were used to describe the dimensions of the helix. Wh represented the initial width of the helix, Aw represented the angle through which the helix wrapped around the central axis of the port and Hs was used to define the maximum height of the helix roof above the port throat. The scheme is shown diagrammatically in Figure 1(b).

C606/001/2002 © IMechE 2002

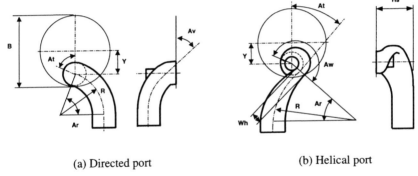

(a) Directed port   (b) Helical port

Figure 1: Port Design Schemes

2.1.4 Final design scheme – parameter ranges

Design guidelines, databases and consideration of the relevant geometry were used to define ranges for each of the parameters. For example, the minimum value of E was determined by considering the likely position of a centrally-mounted fuel injector and surrounding boss. The maximum value was governed by a design guideline recommending the minimum clearance between valve head and cylinder bore. The range for Dn was determined by analysing a database of 4V HSDI engines. Over a range of cylinder bore sizes from 70mm to 100mm, approximately 90% of the engines surveyed had a Dn value between 0.30 and 0.34. Changes in Dn were applied by varying the size of the cylinder bore rather than the valve and port. This made it possible to investigate the relative size of inlet valves whilst maintaining an acceptable number of physical models for the test programme. A 27mm valve inner-seat diameter was chosen and used with cylinder bore sizes between 80mm and 90mm to provide the required Dn range. As non-dimensional flow characteristics were used in the experimental phase of the project, this approach was considered to be valid.

2.2 Experimental Design

A Design-of-Experiments (DoE) approach was adopted to investigate the relationships between design parameters and port performance in a systematic and efficient manner. This approach has been successfully demonstrated in previous studies (8,10), using simplified port geometry and has also been used extensively in other fields of engine research (13,14). Classical experimental plans were chosen to provide robust models suitable for visualisation and optimisation using Response Surface Methods (RSM). The likelihood of non-linear effects and possible interactions between input parameters necessitated the use of these plans in preference to simpler screening experiments. A three-level central composite (CC) plan for six factors was selected for the directed port scheme, requiring 53 test runs. This type of plan was chosen to provide second-order main effects and linear two-way interactions. A three-level Box-Behnken (BB) plan for seven factors was chosen for the helical port scheme; 62 experiments were required, compared to 84 for the equivalent CC plan. The standard run order of both plans was randomised in order to provide a final test matrix. Within each test plan, several repeat tests ("centrepoints") were performed to provide a measure of repeatability.

2.3 CAD model construction and hardware manufacture

The generic design scheme was used to develop parametric 3-D CAD models. Throughout the design process, realistic manufacturing features were added, including a machined throat and a pocket to accept the valve guide. In the helical port design, a valve guide boss was added and all radii were compatible with typical casting requirements. Particular attention was paid to the robustness of the models to ensure that the geometry could be modified using the main driving parameters. The CAD process also involved generating the required number of model variants to populate the DoE test matrix. For the directed port, 13 unique models were necessary. 33 helical port models were required due to the increased number of hardware variables. Variations in the At and E parameters were achieved by changing the position of the port in the cylinder. Therefore, extra models were not required to investigate them.

Physical port models were manufactured using rapid prototyping (RP) techniques. After trials involving several methods, selective laser sintering (SLS) was chosen as the most suitable. The physical properties of the SLS models were found to be acceptable and an accuracy check was performed by comparing the original CAD surface with a set of curves constructed from a 3-D scan of the physical model. Although some distortion of the part occurred during the production process, this was small compared to the intentional variation in geometry resulting from the parameter changes. A skeletal frame consisting of the essential mating surfaces and location points was constructed to house the port models. The throat and entry ends of each model were fitted into recesses machined into the frame to provide positive location. A poppet valve, valve seat insert and valve guide were designed using typical geometry for a modern HSDI diesel engine. The valve guide was mounted concentrically to the valve seat by means of an adjustable location device. A fixture was also made for manually lifting the inlet valve to a known height above the valve seat using a 1mm pitch screw thread.

2.4 Test apparatus and experimental procedure

A conventional steady flow experimental procedure was followed, details of which may be found in previous studies (3, 11). A brief overview of the procedure is provided in the Appendix. A systematic method of preparing each test was developed to ensure that the experimental plan was followed accurately. This involved the use of unique identification labels for each RP model and a pre-test checklist to confirm parameter settings.

2.5 Data analysis

The raw test data was processed using a standard software routine (10) to calculate flow parameters. Non-dimensional rig swirl (Ns) and flow coefficient based on constant inner-seat area (Cf) were calculated at each valve lift point. Mean flow coefficient (MCf), gulp factor (inlet mach index, Z) and swirl ratio (Rs) were also calculated using all test points plus additional engine data and valve lift profile. Raw and processed data for all models were then recorded in a database for further analysis. Definitions for these flow parameters are provided in the Appendix.

The processed test data was analysed using MODDE, a commercial software package from Umetrics AB, to identify significant design parameters and construct polynomial response surface models. These were used to interpret the results and visualise the relationships between the parameters and responses. Alternative models, using either the individual Ns and Cf values at each valve lift, or summary parameters Rs and Z, were developed. The Ns and Cf models were used to investigate the detailed effect of parameters throughout the valve lift range. The Rs and Z models were then used to develop a software program based on Microsoft Excel and Visual Basic to optimise port geometry for a range of realistic

C606/001/2002 © IMechE 2002

performance targets. The port design parameters were used as input variables and the Z response was minimised, subject to various Rs targets and external constraints, to provide trade-off curves and other significant trends.

3   TEST RESULTS AND DISCUSSION

3.1   Directed ports

A summary of the test results for the directed port study is shown in Figure 2. Cf values were highly dependent on valve lift as expected, exhibiting a typical rising trend before flattening off as maximum valve lift was reached. The variation in Cf values was much greater at high lift, indicating that the valve seat geometry governed low lift characteristics. A greater range of Ns values was observed at all valve lifts, although the greatest variation was at high lifts. Ns values at 10mm varied from approximately –0.1 to 0.6, indicating that the design parameters strongly influenced swirl direction and magnitude.

Multiple linear regression was used to determine the most significant parameters and interactions. *Av* (vertical port angle) was the most important parameter in terms of the Cf response throughout the valve lift range. *At* (tangential approach angle), *Ar* (port curve angle) and *R* (port curve radius) were also influential, as were the *Av*Ar* and *Ar*R* interactions. The Ns responses varied with lift but *At*, *Ar*, *E* (eccentricity) and the *At*Ar* and *At*E* interactions were important throughout. *Dn* (non-dimensional valve diameter) did not significantly influence swirl generation or flow performance, although this was partly due to the setting of engine parameters to ensure constant mean inlet gas velocity. Therefore, this result only showed the effect of *Dn* in terms of its impact on the position of the valve in the cylinder. The real effect of *Dn* in engines is likely to be significant due to gas velocity effects; these may be investigated by post-processing of the raw steady-flow results using alternative bore/stroke ratios or engine speeds.

The quality of fit for the DoE models was examined using the squared correlation coefficient R^2. A comparison of observed responses with those predicted from the model was also used as an indication of model quality, as shown in Figure 3. R^2 values tended to increase with valve lift as the parameter effects became more significant and the flow regime became more stable. The regression analysis results were used to investigate the relationships between input parameters and performance responses. The most significant Ns and Cf responses are shown in Figure 4. The predicted Ns response, shown in Figure 4(a) indicated that high valve-lift Ns values increased with *At*. This response is logical when the geometry is considered. High *At* values were associated with increased tangential flow into the cylinder and therefore increased swirl. Low valve-lift Ns values were also increased but to a lesser extent. The effect of *At* on the Cf response is also apparent, as was observed from the regression analysis. Low valve-lift Cf values were not sensitive to port geometry changes, but the high-lift response to *Av* was significant, as shown in Figure 4(b). An increase in *Av* caused an improvement in flow performance at high valve-lifts. This was likely to be due to reduced losses in the transition area where the port blended into the throat. Figure 4 also shows experimental Cf and Ns curves for comparison. The agreement with predicted values was excellent, although this was expected as these test results were used to build the DoE model. The *At* parameter effect was investigated further by performing an additional set of experiments. *At* was varied from minimum to maximum in several steps to characterise the true response and this was compared with the predicted values. The results are shown as a surface plot in Figure 5, indicating a good level of agreement between experimental and predicted values. Note that the predicted *At* effect each 1mm valve lift increment is represented by an independent

C606/001/2002 © IMechE 2002

second-order model and the continuous surface was produced by hermite interpolation. The same treatment was also applied for the observed data, although the At parameter effect does not follow a smooth trend throughout the range of valve lifts and therefore the surface has a "crumpled" appearance.

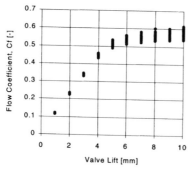 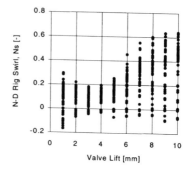

Figure 2: Results summary (directed ports)

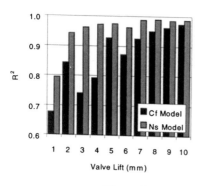 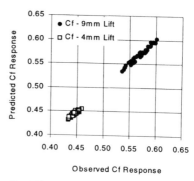

Figure 3: DoE model quality (directed ports)

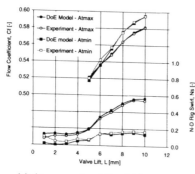 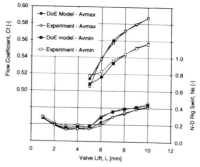

(a) At parameter, Cf and Ns responses

(b) Av parameter, Cf and Ns responses

Figure 4: Performance responses (directed ports)

C606/001/2002 © IMechE 2002

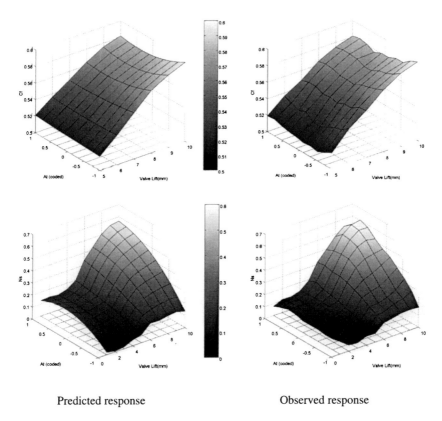

Predicted response   Observed response

Figure 5: *At* validation (directed ports)

3.2 Helical ports

In comparison with the directed port results, swirl levels for the helical ports were generally higher, particularly at lower valve lifts. Flow performance suffered as a result, with generally lower maximum Cf values. The rate of increase of Cf at low-lift remained similar, as expected given the common valve seat geometry. The data was analysed using the same method described for the directed ports. The model fits were comparable to those of the directed port, with approximate R^2 values of 0.94 for the Ns responses and 0.92 for the Cf responses. It was noticeable, however, that the low-lift Ns models were more robust than the equivalent directed port models. This was attributed to the likely presence of a well-defined flow structure, due to the helical port features. Regression analysis showed that the most significant parameters were *At*, *Aw* (helix wrap angle), *Wh* (helix width) and *Hs* (helix start height). The importance of the *At* parameter, shown in Figure 6(a), indicated that the helical ports relied to a certain extent on a tangential flow component to generate swirl. At low *At* settings, maximum Ns was achieved at approximately 6mm lift. As *At* increased, the maximum Ns value occurred at 10mm lift. Large *Wh* values (i.e. a wide helix) resulted in a rapid decrease in swirl as valve lift increased, as shown in Figure 6(b). The loss in high-lift swirl was possibly due to the inlet air travelling further around the helix before exiting into the cylinder. This could result in the main jet flow from the port striking the cylinder wall or creating swirl in a reverse direction with respect to swirl created in the helix. Further analysis is required to fully understand this effect. *Wh* and *Hs* influenced the Cf response significantly (Figures 6(c) and (d)), this was attributed to their effect on the flow area of the critical helix entry section of the port.

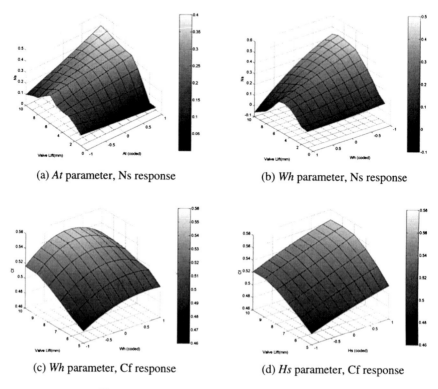

(a) *At* parameter, Ns response

(b) *Wh* parameter, Ns response

(c) *Wh* parameter, Cf response

(d) *Hs* parameter, Cf response

Figure 6: Performance responses (helical ports)

5   KNOWLEDGE-BASED MODEL

The single-port DoE models were used to create a predictive multi-valve model. The Rs and Z response models were used for this purpose. An additive method used by previous workers (4,5) was adopted to estimate the performance of a multi-valve configuration from the individual port flow predictions. A range of configurations could be simulated by combining any two ports and the resulting performance trends investigated. It is important to note that, although this approach was limited as flow interactions between ports could not be characterised, its use has been demonstrated successfully. During the concept stage of the design process, the primary requirement is to evaluate candidate solutions rapidly and compare them objectively. An optimisation routine was developed in Microsoft Excel to maximise the flow performance of a particular port configuration for a given swirl ratio

C606/001/2002 © IMechE 2002

target. Predicted performance for three typical port configurations is shown in Figure 7. Combinations using either two directed ports, two helical ports or one of each type were optimised to assess their operating ranges. Twin directed configurations were most efficient at low swirl ratios. Twin helical configuration were capable of higher swirl ratios but generated low swirl less efficiently. Mixed configurations performed well over an intermediate swirl range, but were also less efficient than the directed ports at low swirl ratios. The results showed that it is important to consider the likely swirl requirements of the combustion system during the concept design phase in order to select the most appropriate port configuration. Experimental results for these configurations are also shown in Figure 7. The experimental results were obtained using new, multi-valve RP models. The accuracy of predictions was encouraging, indicating that the multi-valve model performed well. However, further tests are required to fully validate the model and identify conditions under which the additive assumption is sufficient.

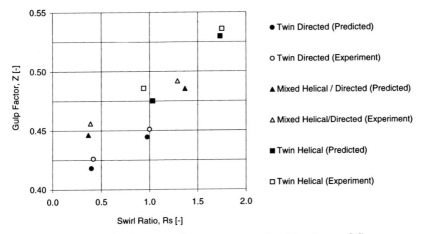

Figure 7: Optimum performance curves (multi-valve model)

6   CONCLUSIONS

- A knowledge-based model of inlet port performance for HSDI diesel engines has been successfully developed
- Robust generic port models have been constructed to capture the major design parameters of helical and directed inlet ports and a design-of-experiments approach has been employed to quantitatively assess their influence on port performance characteristics
- A range of typical multi-valve design concepts have been investigated and port performance optimised whilst meeting realistic external constraints

ACKNOWLEDGEMENTS

The authors would like to thank the directors of Ricardo Consulting Engineers Ltd. for their permission to publish this paper. The authors would also like to thank their colleagues at

Ricardo Consulting Engineers and the University of Brighton for their valued contributions throughout the project. Support from The Royal Commission for the Exhibition of 1851 is also gratefully acknowledged.

REFERENCES

1. I P Gilbert, A R Heath and ID Johnstone. Multivalve High Speed Direct Injection Diesel Engines – The Design Challenge. ImechE Seminar on Recent Advances in the Mechanical Design and Development of Engines and their Components. London, 1992
2. N F Gale. Diesel Engine Cylinder Head Design – The Compromises and the Techniques. SAE900133
3. M L Monaghan, H F Pettifer. Air Motion and its Effects on Diesel Engine Performance and Emissions. SAE810255
4. J Kawashima, H Ogawa, Y Tsuru. Research on a Variable Swirl Intake Port for 4-Valve High-Speed DI Diesel Engines. SAE982680
5. Y Li, L Li, J Xu, X Gong, S Liu, S Xu. Effects of Combination and Orientation of Intake Ports on Swirl Motion in Four-Valve DI Diesel Engines. SAE2000-01-1823
6. S K Widener. Parametric Design of Helical Intake Ports. SAE950818
7. H Blaxill, J Downing, J Seabrook, M Fry. A Parametric Approach to Spark-Ignition Engine Inlet-Port Design. SAE1999-01-0555
8. A Brignall, Z M Jin. Investigation of Inlet Port Design on Engine Swirl using Orthoganol Array Experimentation and Computational Fluid Dynamics. ImechE Conference on Fluid Mechanics and Dynamics of Multi-Valve Engines. London, 9 June 1999
9. V J Page, G D Blundell. The Application of Design of Experiments and CFD to the Optimisation of a Fully Parametric Helical Port Design. Computational and Experimental Methods in Reciprocating Engines, London, Nov 2000. ImechE Paper C587/014/2000
10. M C Bates, M R Heikal. A Knowledge-Based Model for Multi-Valve Diesel Engine Inlet Port Design. SAE 2002-01-1747
11. P Lambert. Computer Applications User Note: Blowrig. DP 94/2148 (Confidential)
12. C D De Boer, R J R Johns, D W Grigg, B M Train, I Denbratt, J R Linna. Refinement with Performance and Economy for 4 Valve Automotive Engines. Automotive Power Systems - Environment and Conservation. ImechE Paper C394/053, 1990
13. S P Edwards, D P Clarke, A D Pilley. The Role of Statistics in the Engine Development Process. Statistics for Engine Optimisation Seminar, IMechE, London, 2 Dec 1998
14. R A Bates, R Gilliver, A Hughes, T Shahin, S Sivaloganathan, H P Wynn. Fast Emulation of Mechanical Designs Using Computer Aided Design/Computer Aided Engineering Emulation: A Case Study. Journal of Automobile Engineering, IMechE, Vol 213, no D1, Part D, pp27-35

APPENDIX

EXPERIMENTAL PROCEDURE

Prior to each test, the impulse swirl meter (ISM) was checked, zeroed and spanned to ensure a linear response over the operating range. The pressure at the port inlet was set according to the valve size to ensure turbulent flow in the port. The model assembly was visually checked, then tested for air leaks by closing the valve and running the fan to a pressure above the required test pressure. A measured flow through the laminar flow meter (LFM) would signify a leak between it and the port model. Any small leaks were sealed before commencing the first test. The inlet valve was opened 1mm above the valve seat using the lifting screw, with the fan running. Gauge pressure at the port entry was then set by adjustment of the fan speed. Following a short period to allow the instruments to settle, the test results were recorded. A series of test points were recorded with the inlet valve lift increased incrementally by 1mm to a maximum beyond that expected in an operating engine. A maximum lift of 10mm was deemed sufficient in this study. At each test point, the fan speed was adjusted to maintain a constant gauge pressure at the port entry.

Equations used in steady flow analysis:

$$Cf = \frac{Q}{AV_o}$$

Q = volumetric flow rate (m^3/s)

A = reference area (inlet valve inner seat area)

$$V_o = \sqrt{2\delta P/\rho}$$

(velocity head)

ΔP = differential pressure across port and valve (Pa)

ρ = inlet air density (kg/m^3)

$$Ns = \frac{8G}{\dot{m} B V_o}$$

G = swirl meter torque (Nm)

m = air mass flow rate through port (kg/s)

B = cylinder bore (m)

$$Z = \left(\frac{B}{D}\right)^2 \frac{2S\omega_E}{na\,MCf}$$

D = inlet valve inner seat diameter (mm)

ω_E = engine crankshaft speed (rad/s)

a = sonic velocity at port inlet conditions

$$MCf = \frac{\int_{\alpha1}^{\alpha2} Cf\, d\alpha}{\alpha2 - \alpha1}$$

$\alpha1, \alpha2$ = crank angle, inlet valve opening and closing

$$Rs = \frac{\dfrac{BS}{nD^2}\int_{\alpha1}^{\alpha2} Cf\, Ns\, d\alpha}{\left[\int_{\alpha1}^{\alpha2} Cf\, d\alpha\right]^2}$$

C606/032/2002

Practical application of DoE methods in the development of production internal combustion engines

K ROEPKE, A ROSENEK, and M FISCHER
IAV GmbH, Berlin, Germany

ABSTRACT

This article describes the practical application of DoE methods in the development of production internal-combustion engines. Examples demonstrate how application of DoE can simplify production engine development. This paper also discusses practical problems encountered in the application of DoE, and describes possibilities toward their solution.

1 INTRODUCTION

Modern engines contain a great number of controlled components as well as powerful electronic control units. These systems enable conformity with the many and various regulations imposed by strict legislation – and satisfy users' expectations as well. For engine development, however, this entails a noticeable increase in expenditure of resources to determine the optimal combination of all selectable parameters. The volume of data input into electronic control units resulting from these developments has appreciably risen in recent years.

Table 1: Requirements placed on conduct of testing

• Exact measurements:	– Systematic preparation of testing – High measuring accuracy and precision
• Faster measurements:	– Automation of the required measuring functions – Robust test processes
• Reduced extent of measurement:	– Reduction in test program extent by application of optimization and simulation functions – Low rates of test repetition

Since the number of measuring points required for complete operational-test measurements rises exponentially with the number of input variables, it is obvious that it is possible to conduct map-based operational measurements and manual evaluation for only selected influencing variables. In addition to complex operational-test measurements required for high-dimensional maps, further restrictions come into play. Constraints on product development schedules and the resulting restriction in available time, as well as limitations in development funds, have produced demands for reduction of measurement time and expense – but without reduction in quality of results obtained. These circumstances have resulted in the following requirements for the conduct of testing (table 1).

Fulfilling these specifications is an essential prerequisite for obtaining sufficiently exact results while conforming with the stated requirements for reduced extent of measurement. The present publication will show how techniques of design of experiments (DoE) (8) can enable significant reduction in specific development tasks for internal combustion engines. Three examples will demonstrate the practical application of these methods.

A number of investigations have treated the application of DoE in engine development. Until now, however, broad DoE application has not taken place. One of the chief obstacles to wider implementation is the requirement for testing engineers to modify working techniques. To achieve optimal benefits from DoE, test engineers must in fact discard the conventional trial-and-error approach as their conventional work method, and replace it by the following sequence: planning, measuring, checking, and simulating. In addition, test engineers have traditionally shown hesitancy with respect to simulation. By application of actual measured values, and by eliminating physical assumptions, however, DoE eliminates the chief points of criticism leveled at classical simulation. Support from competent DoE consultants, as well as intensive interaction in teamwork, can consequently enable rapid transformations in the way test engineers work (3, 4).

2 VARIABLE CAMSHAFT PHASING

Although substantial information is known on the fundamental effects exerted by engine speed and cylinder charging on valve timing optimal for fuel consumption and torque, it is still necessary to determine by extensive bench testing the control times individual to a particular engine.   The result is a high-dimension parameter space with numerous interactions: this is because, for example, the ignition angle optimal for fuel consumption, and the knocking limit, will depend on variable camshaft phasing. In cases of complete measurement of these interrelationships, several tens of thousands of measurements may easily prove necessary. The results are available in the form of individual characteristic curves. It is very difficult, however, to perform optimization procedures intended for several target variables. Modifying the optimization criteria can likewise lead to complicated and tedious analysis of large masses of data.   This would be necessary, for example, upon changing the criterion from design for optimal fuel consumption, to design allowing a combination of optimal consumption and optimal emission. Here as well, application of DoE can lead to appreciable reduction in measurement effort, and to simplified evaluation procedures (1, 2).

Fewer than 1,000 measurements are necessary when using DoE in building models for torque, NO_X and HC emissions, COV (coefficient of variation of IMEP), knocking limit, and catalyst temperature. The input factors for these models are speed, load, ignition angle, intake and exhaust camshaft phasing, and air-fuel ratio.

In the present study we designed our experiment using the D-optimal algorithm. To enhance model accuracy, we divided the operating range of the engine into smaller parts, for each of which we performed modeling. As result, there are several models for every operating range

C606/032/2002 © IMechE 2002

and response. The models can be used to define optimal valve timings and ignition points, and it is also possible to use them for input of data for the torque interface.

Figs. 1 and 2 show model values for one operating point (constant engine speed and constant cylinder filling), as a function of camshaft phasing: for effective fuel consumption – with respect to NOx and HC emissions – and for the ignition angle. It is evident that large values for valve-opening overlap are associated with a reduction not only in fuel consumption, but also in emissions.

Fig. 1 shows a calibration for minimal fuel consumption. The optimal settings for camshaft phasing are marked in the maps for fuel consumption, ignition point, and emissions. Sacrificing 2% of fuel consumption will allow even further emission reduction. Figure 2 shows calibration for minimal emissions with 2% loss of fuel consumption in comparison to minimal fuel consumption. These calibrations can be calculated from the models, without requirement of further measurement.

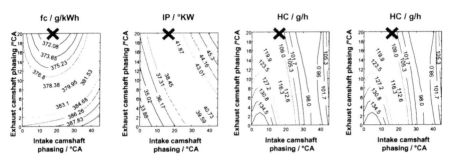

✘ $Optimum = \min(f_c)$

Figure 1: Calibration for minimal fuel consumption

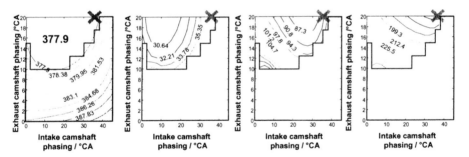

✘ $Optimum = \min(HC + NO_x), f_{c,min} + 2\%$

Figure 2: Calibration for minimal emissions by sacrificing 2% fuel consumption.

3 DIRECT-INJECTION SPARK-IGNITION ENGINES

In comparison to engines with intake-manifold fuel injection, DISI engines have a number of additional degrees of freedom. In homogenous operation, for example, it is also necessary to optimize the EGR rate, the charge-motion valve (if present), as well as the injection initiation point – in addition to the ignition angle and variable camshaft phasing. In homogenous-lean fuel operation, the fuel/air ratio must furthermore be optimized. Figure 3 shows some DoE-modeled values in dependence of ignition point and EGR. The area of validity of models is marked.

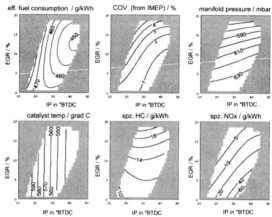

Fig. 3: Fuel consumption, manifold pressure, variation coefficient, exhaust-gas temperature, and emissions in homogenous operation with a DISI engine.

In stratification operation, it is necessary to optimize the following in addition to the ignition angle, the EGR rate, and the variable camshaft phasing: the end-point of fuel injection, rail pressure, as well as manifold pressure. A stable ignition-angle/end-of-injection window, which can often be very narrow, is a problematic factor in application. It can be seen in figure 4 that he area of validity, which is here the stable window is much smaler than in homogonous operation.

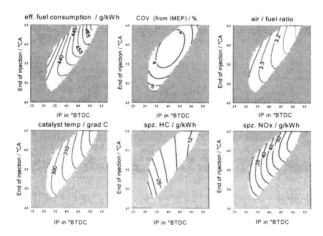

Fig. 4: Fuel consumption, air ratio, variation coefficient, exhaust-gas temperature, and emissions in stratification operation of a DISI engine.

                                                  C606/032/2002 © IMechE 2002

4 COMMON-RAIL DIESEL ENGINES

For common-rail diesel engines, the number of degrees of freedom is likewise very large; in addition, this number will in future continue to increase. Today, the standard includes pilot injection and main injection, in addition to variables for air mass and supercharging pressure. Systems with up to five injections are tested, for which it is necessary to determine the point in time for injection, in addition to the individual quantities. Here as well, DoE methods are practically ideal for determining the manner of operation and interrelationships of these variables with respect to fuel consumption, noise, and emissions (5). Fig. 5 shows the individual effects on the exhaust-gas components NO_x and soot, as well as on noise and fuel consumption. In addition to the classical conflict among application objectives, these representation show various degrees of influence of the parameters on the target variables.

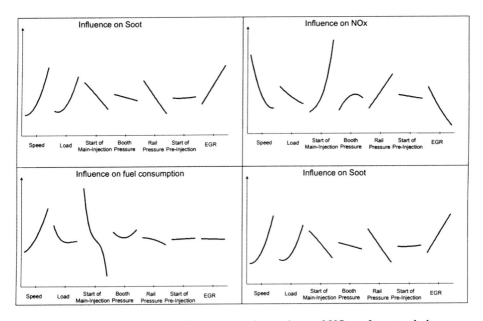

Fig. 5: Individual effects on fuel consumption, noise, and NO_x and soot emission.

5 PROBLEMS IN PRACTICAL APPLICATION OF DOE-METHODS

In this paper we do not intend to ignore the problems that arise in practical application of DoE methods:
For optimal design of experiments, the experiment space must be known. This knowledge is required to distribute the measurements throughout the experiment space such that it is well covered. In actual practice, however, this knowledge is often not available, especially in case with many degrees of freedom. In such events, extensive preliminary investigations are required in order to determine the actual experiment space. In addition, the settings for the individual measurements differ so greatly from each other that manual test measurements take a great deal of time and are often subject to error. The solution to these problems is

intelligent automation of the test bench: automation that allows not only various measurement strategies, but also differentiated reactions to violations of limit values. IAV has developed an in-house solution which determines the experiment spaces automatically at the test bench: which in turn enables creation and measurement for optimal, realistic design of experiments. This procedure enables enhanced model quality, with time savings from automation at the same time. Figure 6 shows an example screen shoot of an automatic measurement of a 9 parameter diesel engine.

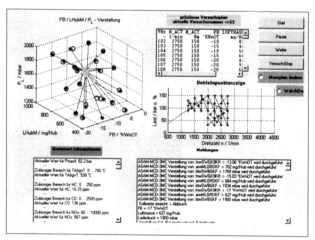

Fig. 6: Intelligent test-bench automation

The results of DoE work are mathematical models that are used for analysis. As long as only DoE experts were able to perform this analysis, test engineers – who are more practically oriented – would not accept DoE technology. For this reason, IAV developed its MODEL ANALYZER, which enables simple performance of extensive model analyses. It is possible, for example, to show a number of two- or three-dimensional graphical presentations, to prepare individual-influence diagrams, to perform robustness experiments, and to carry out various optimization processes (figure7).

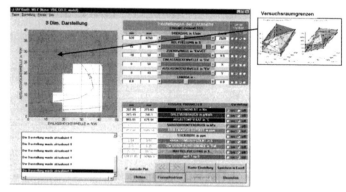

Fig. 7: The IAV MODEL ANALYZER, for simple analysis of DoE models

     C606/032/2002 © IMechE 2002

6 DOE PHASE MODEL FOR DYNAMIC MODELING

Until now, the methods of statistical DoE have been primarily applied for the optimization of static states. For emission optimization in dynamic engine operating states (8), IAV employs the so-called DoE Phase Model.

Fig. 8 shows the plot of pollutant emission from a spark-ignition engine after cold starts: here, for the New European Driving Cycle (NEDC). It is evident that approx. 90% of total HC and CO emissions shown for spark-ignition engines takes place within the first 80 s. This is because a catalytic converter requires a certain time to reach conversion temperature after a cold start. The converter reaches conversion temperature when the exhaust-gas temperature downstream of the converter exceeds the upstream temperature.

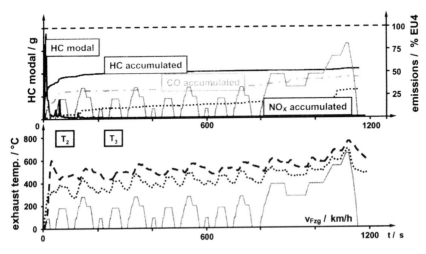

**Fig. 8: Emissions for spark-ignition engines in the
New European Driving Cycle (NEDC)**

From these circumstances it is possible to derive two general objectives of exhaust-gas applications for spark-ignition engines. The first objective is to effectively apply control-technology strategies implemented in the ECU, to enable the catalytic converter to heat up as fast as possible, and to quickly achieve optimal conversion rates. The second objective is to ensure that the start and post-start phases are coordinated so as to achieve optimal conditions for exhaust-gas emission upstream of the catalytic converter.

In our work with the DoE Phase Model, the first 80 s after engine start were divided into meaningful individual phases, and then developed a stady-state DoE model individually for each of these phases (see Fig. 9). A d-optimal design was used for the DoE. This procedure offers the advantage of making it possible to consider the most effective variation parameters for each individual phase. The heating measures implemented for the catalytic converter do not modify the statically conducted input of data into controllers for the parameters of the ECU: i.e., ignition angle, mixture, and cylinder filling. As a result, it is possible to individually calibrate each phase in the dynamic phases by means of input of block data into the controller, or simple input of gradient data. With respect to the individual phases, this procedures makes it possible to know the initial and (after analysis) the final conditions.

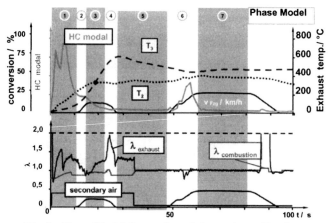

Fig. 9: Phase Model for static and dynamic applications

This procedure further provides all transition conditions for the overall investigation of all phases, and enables a self-contained, unified study approach. The target parameters for the overall DoE are total emissions, as well as the data for plot of the exhaust-gas temperatures upstream and downstream of the catalytic converter.

The example of secondary-air exothermy for fast heat-up of the catalytic converter clearly reveals that a different optimized map-based pilot control for the air-fuel mixture is necessary: for the λ combustion chamber alone.

7 DYNAMIC CALIBRATION

As stated earlier, the application of DoE appreciably reduces the number of experiments and, in turn, the time required for development.

On the basis of long years of experience gained at IAV in the use of dynamic engine test benches, and in the employment of DoE in many areas of engine development, we have for some time now carried out development of exhaust-gas concepts on the basis of a combination of these two systems. On the dynamic engine test bench, external forced cooling of cooling water, engine oil, intake air, and exhaust system has drastically reduced the traditionally long vehicle conditioning periods (6 … 12 h until recently). As a result, it is now possible to start a new cycle after less than 2 hours. IAV employs two methods, with their respective application according to specific testing task, for transfer of driving cycles to the dynamic engine test bench.

The first method involves a transfer technique that, as an initial step, records the required driving cycle with an actual test vehicle and driver on the exhaust roller dynamometer test bench. This first technique then, as second step, involves transfer of the cycle to the engine test bench. The second method uses special software to completely simulate both the vehicle and the driver: which means that an actual vehicle is not necessary (see Fig. 10). The comparisons in cycle FTP-75 (see Fig. 11) demonstrate the quality of vehicle simulation on the test bench.

C606/032/2002 © IMechE 2002

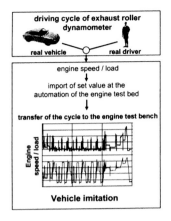
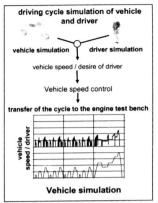

Fig. 10: Vehicle model and vehicle simulation on a dynamic engine test bench

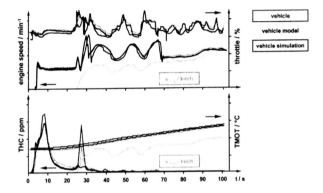

Fig. 11: Comparison of vehicle, vehicle model, and vehicle simulation

The comparability of engine-speed data plays a decisive role in transfer of the driving cycle from the exhaust roller dynamometer test bench to the engine test bench. Since direct torque comparison between vehicle and engine test bench is possible only at prohibitive expense, load comparison takes place by using the throttle-valve angle. The most important target parameters for the exhaust application specialist is the comparability of data for emissions, exhaust-gas temperatures, and mixture ratio.

The progression of temperature changes in oil and in coolant have significant influence on emission behavior, since parameters for oil and coolant temperature represent the input parameters for various maps within the engine electronic control system. Temperature differences in the coolant cause great fluctuations in emissions, especially during the warm-up phase.

During the concept phase, the reproducibility of measurement is of critical importance, since current low-emission concepts make it absolutely necessary to identify the slightest emission differences for purposes of evaluation and analysis. Fig. 12 clearly shows the slight deviations with respect to vehicle measurement, and the high degree of reproducibility of the individual measurements.

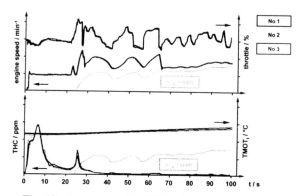

Fig. 12: Cycle reproducibility of the vehicle model

8 AUTOMATED STANDARD ANALYSIS AT IAV

Despite great reductions in the effort required for experiments, the conduct of testing, evaluation work, and detailed analysis of measured data still consume the majority of time required. During this crucial phase of work, it is essential to promptly and exactly determine any measurement faults to enable any necessary repetition of measurements. To enable fast and objective-oriented analysis, IAV applies an extensively automated analysis tool: only 10 minutes after the exhaust-gas test, the test engineer has a complete evaluation of all measured signals (See fig. 13).

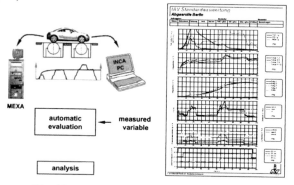

Fig. 13: Automated standard analysis at IAV

The DoE expert transfers the measured data into the DoE model and provides the test engineer a "best-of" data status. Experience gained from various projects has revealed that such a data status can under certain conditions prove very difficult to realize, given the functionality of current ECUs. Data acquisition of evaluations of DoE results in *n*-dimensional space, furthermore, can likewise prove very difficult. It is for these reasons that IAV has developed its Model Analyzer, as stated above. The Model Analyzer affords the test engineer the possibility of implementing various applications offline, with consideration

C606/032/2002 © IMechE 2002

taken, for example, of driver-behavior aspects. The system also provides him with calculated emission plots.

With the aid of the DoE methods described above, in combination with a dynamic engine test bench, IAV has – on the basis of ULEV exhaust-gas coordination – optimized exhaust-gas application for a series-production four-cylinder spark-ignition engine with exhaust-gas turbocharging. Optimization of exhaust-gas emission upstream of the catalytic converter, together with implemented strategy of catalytic-converter heating by application of DoE, resulted in reduction of HC emissions by approx. 11% with respect to ULEV limits. Results accordingly fell approx. 15% below HC SULEV limits, without additional hardware modifications with a new catalytic converter (see Fig. 14). NO_x emissions were within the context of this project.

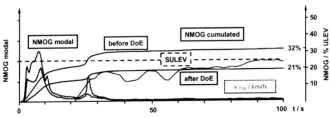

Fig. 14: Optimization results

9 CONCLUSIONS

Statistical methods such as DoE can accelerate and enhance the development process. A prerequisite for this improvement, however, is solution of practical problems in DoE application. An intelligent automation strategy, for example, must be possible at the test bench, in order to produce effective and high-quality experiment designs and models. Tools are additionally required, in order to enable fast model evaluation by non-experts as well. This paper elaborates on the benefits of the DoE method in examples from development of production spark-ignition engines with intake-manifold fuel injection, direct-injection spark-ignition engines, and common-rail diesel engines, and from application in dynamic engine application.

10 REFERENCES

1. Roepke, K.; Fischer, M.: Efficient Layout and Calibration of Variable Valve Trains, SAE2001-01-0668
2. Seabrook, J.; Nightingale, C.; Richardson, S. : The Effect on Hydrocarbon Emissions – An Investigation with Statistical Experiment Design and Fast Response FID Measurements. SAE 961951
3. Mitterer, A.; Fleischhauer, T.; Zuber-Goos, F, Weicker, K.: „Modellgestützte Kennfeldoptimierung an Verbrennungsmotoren". VDI Berichte 1470, Mess- und Versuchstechnik im Fahrzeugbau, Kongress 29./30.4.1999 in Mainz, S. 21-36
4. Fischer, M.; Röpke, K.: Effiziente Applikation von Motorsteuerungsfunktionen für Ottomotoren. MTZ 61/2000, S.562-570
5. Bredenbeck, J.; Fehl, G.; Dohmen, H.-P.: Verkürzung der Entwicklungszeiten für moderne Antriebe durch VEGA. ATZ/MTZ Sonderausgabe: System Partners 1997
6. Geschweitl, K.; Pflügel, H.; Riel, A.; Leithgöb, R.: Cameo in Online Engine Optimization Using Statistical Methods, 4. int. Symposium on Internal combustion diagnosis, Baden-Baden. 2000
7. Lipson,C.; Sheth, N.J.: Statistical Design and Analysis of Engineering Experiments, New York, 1973
8. Andrews, M; Cary, M.; Hilton, D.: The Application of Design of Experiments and Multi Variable Optimization to Cold Starting in an Automotive SI Engine. SAE 981413

C606/013/2002

A model and methodology used to assess the robustness of vehicle emissions and fuel economy characteristics

P J SHAYLER and **P I DOW**
Department of Mechanical Engineering, University of Nottingham, UK
M T DAVIES
Ford Motor Company, Laindon, UK

ABSTRACT

A suite of models (NuSim) allowing vehicle fuel economy and emissions performance to be simulated is described. This is coded in Matlab Simulink and will run in near to real-time on high-specification personal computers. The application of NuSim to assess the robustness of fuel consumption and tailpipe emissions to uncontrolled variations in design factors is described. A DOE array covering 16 factors at 3 levels and requiring 291 simulated NEDC test runs formed the basis for the assessment. The study illustrates the practicability of investigating robustness and quality assurance through simulation and the use of a DOE methodology.

1. INTRODUCTION

The automotive industry is engaged in a continuing drive for product improvement, shorter development cycles and lower development costs. Motivating factors include competition for customers, profitability and return on capital, and the need to meet mandatory standards of vehicle performance, most notably those concerned with limiting the emissions of pollutants from vehicles. These various pressures make it essential that, as far as possible, the vehicle development process is informed by knowledge of how the design will perform in service. It is now becoming practicable to generate this knowledge using engine and vehicle performance simulation software. In prospect, this will profoundly change the development process to one in which the influence of detail on total vehicle behaviour can be weighed from the earliest stages of design specification. There is, however, still much to learn about how and what can be simulated, how this capability can best be utilised, and what part the statistical and analytical methods might play. The aim here is to illustrate recent developments in this area by reference to an example study. The study concerned the robustness of fuel consumption and tailpipe emissions of a new mid-size saloon to uncontrolled variations in the behaviour characteristics of sensors and actuators, and selected other features of the engine design. To the authors' knowledge, this was the first study of its type. The study was made possible by the development at Nottingham University of a set of systems and process models. The integrated set provides a total powertrain and vehicle model (NuSim). At the time of building the model version for this particular application, pre-production vehicles were running. Engine Management System (EMS) software was still being developed and refined, and minor hardware changes were still being evaluated. Limited test bed data were available on steady-state engine performance and emissions, as were the results of first New European Drive Cycle (NEDC) tests, by the time the model was sufficiently complete to evaluate. Thus the development and evaluation of the model was carried out in parallel with the later stages of the vehicle development programme. At this point, it became practicable to carry out the study of robustness using the model. It would have been prohibitively costly and probably impracticable to have carried out the same study through vehicle testing. The paper covers a brief overview of the NuSim and an outline of the robustness study, the methodology and use of Design of Experiments (DOE) in this, and illustrations of outcomes.

2.   STRUCTURE AND OVERVIEW OF NUSIM

The forces driving growth in the use of simulation tools have been described by Moskwa et al [1], who also outlined a philosophy for the formulation of models suited to the application. NuSim has been developed for a number of uses. Illustrated here are the prediction of vehicle fuel economy and emissions, and studies of performance robustness to variations in component and sensor characteristics. Other uses include the desk-top development of control strategies and calibrations, and the evaluation of OBD and adaptive control algorithms. NuSim is comprised of a set of sub-models for the various systems and processes which define vehicle performance, together with a 'driver' model and specification of target patterns of operating conditions. Commonly, the requirement will be for a simulation of performance over the cycle used for emissions and fuel consumption tests. In the EU, the test cycle is a cold-started NEDC. The model has been written in Matlab Simulink and from the outset was targeted to run on high specification Personal Computers so as to be accessible to a broad and distributed user-base. The requirements of computational speed and accuracy of the model are met by using a mix of physics-based and empirical sub-models. Importantly, the model is designed to incorporate Engine Management System (EMS) software supplied as, or converted into, Simulink code. This feature requires the various timing triggers, inputs from sensors and outputs to actuators from the EMS to be defined and connections made to the plant model. The EMS software is probably the most version-specific and least generic part of NuSim. As far as possible, other parts have been designed to be easily adapted for applications to other powertrains and vehicles. The approach taken defines a small set of generic types covering the major distinguishing features (in-line and V type cylinder layouts, diesel and gasoline, port and direct fuel injection, naturally aspirated and turbocharged). The most appropriate of these generic types is then adapted to create a new version for a particular application.

In the main, NuSim has been organised as a set of systems models with obvious physical boundaries. Thus for the engine, these include the intake system, the combustion system and the exhaust and aftertreatment system. Heat flows and frictional dissipation within the engine are processes simulated using a thermal behaviour model. Each of the models describing system and process features is built individually and then integrated to produce the version of NuSim required. Information passes between the models in the highest layer of the model structure. Successively lower layers reveal detail and provide access to more features of the various model parts. The organisation is illustrated in Figure 1. The most detailed descriptions of behaviour are associated with the plant model of the engine. A version for a

C606/013/2002 © IMechE 2002

four cylinder, naturally aspirated PFI spark ignition engines is used here. Within the plant model, there are eight main submodels simulating systems behaviour. Behaviour and processes are described by physics-based or semi-empirical models. The outputs include state variables such as pressures, temperatures and emissions concentrations. The EMS control strategy and calibration are converted from the C-code format supplied into Simulink for importing into NuSim. The converted code provides a complete version of the strategy. The EMS is linked computationally to the other systems models via connections to models of engine sensors and actuators. The sensor characteristics are defined from supplier data and used to convert sensed physical variables into a simulated voltage signal to the EMS. The EMS control software is run in simulated real time, with operations triggered by timers and trigger pulse trains generated by the relevant sensor representations in the plant model. The sensor signals are sampled at the crank angle or time intervals as actually occurs in the hardware. This mimics the combination of event-based and time-based samplings. The engine plant simulation also describes some processes as being event based, but within a computational scheme which is time based.

The formulations of several of the submodels in NuSim are based on assumptions of ideal behaviour and empirical corrections. In the induction system model, this approach is used to define the flow characteristics of the throttle and valves controlling air bypass, exhaust gas recirculation and EVAP system purge flows, together with the variation of the cylinder volumetric efficiency. The treatment of the last of these can be taken as a typical. For a four stroke engine, cylinder volumetric efficiency is defined by

$$\eta_{cvol} = 2\dot{m}_{gas} / \rho_m V_s N \qquad (1)$$

where $\dot{m}_{gas}$ is the mass flow rate of induced gases, V_s is the engine swept volume, N is engine speed in revolutions per second, and ρ_m is the gas density evaluated at intake manifold conditions. In NuSim, this volumetric efficiency is computed by correcting an approximate, ideal value using

$$\frac{\eta_{vol}}{\eta_{vol,ideal}} = \left[\frac{\sqrt{T_m}\left(T_{c,ref} + 1111\right)}{\sqrt{T_{m,ref}}\left(T_c + 1111\right)} \right] fn_a(N) fn_b(P_m) \qquad (2)$$

The 'ideal' value is given by the expression for an idealised four-stroke induction process derived by Taylor [1],

$$\eta_{cvol,ideal} = \frac{1 + \gamma(r_c - 1) - \left(\dfrac{P_{ex}}{P_m}\right)}{\gamma(r_c - 1)} \qquad (3)$$

where γ is the ratio of specific heats, r_c is compression ratio and (P_{ex}/P_m) is the exhaust to intake manifold pressure ratio. The first of the corrections in equation (2) is for ambient and coolant temperature effects. The form of correction, and the empirical constant (1111), are derived from experimental results given in [2]. The correction is normalised to be unity at a reference ambient temperature T_m of 298K and an engine coolant temperature T_c

corresponding to the thermostat value. The second, $fn_a(N)$, and third, $fn_b(P_m)$, multipliers are engine-specific corrections which reflect the influence of valve timing, intake system geometry, etc. These specific corrections are defined using experimental data in the model version employed here. Alternatively, corrections could have been defined to match predictions of volumetric efficiency made using a wave dynamics code such as Ricardo WAVE. For the throttle, flow through this is described by the equations for compressible flow through an orifice of variable area A defined by throttle position, and an empirical correction abrogating the effects of the assumptions and throttle discharge coefficients. The volumetric efficiency, throttle and valve flow descriptions define boundary conditions for the application of mass continuity and energy conservation to the intake system. These form a mean-value, filling and emptying model of gas behaviour in the intake. The behaviour of fuel injected in the intake port is described by the τ-x model [3] and together with the intake filling and emptying model these define the condition of the mixture induced.

A similar approach is used to define the indicated performance characteristics of the combustion system. The outputs of interest are indicated specific fuel consumption and engine-out emissions of HC, CO and NO_x. Several ways of relating these to indicated operating conditions have been explored [4], [5]. The choice depends on the existence, or otherwise, of test data and the extent and quality of this. In the study reported here, for which a quite substantial data set existed, neural networks were used. The input variables (fuel induced, mixture air/fuel ratio, gas/fuel ratio, spark timing relative to MBT settings, and compression ratio) form part of the set of internal variables within the model. Under cold operating conditions, a semi-empirical model [6] is used to predict fuel losses to the crankcase, and the affects of wall wetting and poor mixture preparation on emissions of HC and CO. As warm-up progresses, these affects diminish to match fully-warm characteristics.

Descriptions of conditions in the exhaust and aftertreatment system cover three aspects of behaviour. The first deals with the need to define exhaust back pressure at the exhaust port, because of its influence on cylinder indicated conditions. The back pressure is determined from the mean-value gas flow rate and a composite resistance coefficient for the system [4]. The second deals with the prediction of mixture conditions sensed by HEGO and/or UEGO sensors. These conditions are predicted using a displace-and-mix model [7] of gas flows from each engine cylinder through the exhaust manifold and pipework down to the location of the sensor or sensors. Finally, a model (PROMEX) is used to predict exhaust gas state throughout the exhaust system, together with a sub-model of the three-way-catalyst, to simulate light-off and the efficiency of emissions conversion for the transient conditions of the NEDC test.

PROMEX is one of the two codes used to simulate thermal behaviour. The second is PROMETS, which provides a lumped capacity thermal analysis of temperature and heat flow distributions within the engine structure, gas-side heat transfer and rubbing frictions from cold start through to fully-warm operating conditions. Both PROMEX and PROMETS have been described in detail elsewhere [8], [9]. A third existing code, FVSMOD [10], has been converted into Simulink code and interfaced with NuSim to simulate the behaviour of the evaporative emissions system.

The transmission and vehicle models are relatively simple but adequate for applications concerned with fuel consumption and emissions characteristics. The flywheel through to the driven wheels is treated as a rigid mechanical system. The driveline losses are represented by

C606/013/2002 © IMechE 2002

a transmission efficiency, which can be defined as a function of gearbox oil temperature, output shaft speed and gear ratio. In the current study, the vehicle specification had a manual transmission for which the gear change points in NEDC simulation runs are defined. Power take-off and gear change characteristics are modelled as will be described in the main text. The vehicle mass, and rolling and aerodynamic resistance characteristics are used to define tractive loads.

The vehicle is 'driven' in simulation through the control of throttle positions to follow a target variation in vehicle speed. A PID controller is used with a feedback correction based on the error between target and actual vehicle speed. The controller can be tuned to reflect a range of 'driver' styles. Solutions are obtained by time-marching, in the time domain. Physical states are updated at typically 2ms simulated time intervals. Internal timers and a simulated crank position input control the execution of the EMS strategy. Thus the EMS operates in both time and event based domains, on tasks carried out at intervals down to $90°CA$.

The accuracy of simulations is illustrated in Figures 2 and 3. The level of agreement between simulation results and data taken from vehicle tests is typical. Predicted and measured induced air mass flow rates over the last section of the NEDC are shown in Figure 2. Comparisons for engine-out and tailpipe NO_x emissions over the NEDC are shown in Figure 3. Similar comparisons have been made for several dozen of the more important EMS software and plant model variables. Even so, this is a small proportion of the total number and the process should perhaps be described as an evaluation based on a sample of variables rather than a validation. The accuracy of individual sub-models within NuSim reflects the use of test data to define characteristics such as engine cylinder volumetric efficiency and throttle discharge coefficient, and to define the relationship between indicated performance and cylinder charge conditions. This constrains the magnitude of errors in simulated behaviour. It also allows a relatively simple and computationally efficiency treatment to be pursued.

3.   ROBUSTNESS STUDY

Fuel economy and tailpipe emissions are two key attributes of vehicle performance and the main objective was to establish how variations in the vehicle build or condition might influence these. Although prototype vehicles were running at the time the study was carried out, it was the availability of NuSim which allowed the robustness of these target attributes to be examined. In part, the purpose was to assess the consequences of uncertainties and provide a measure of confidence that compliance requirements would be met throughout the lifetime of the vehicle. Beyond this, the study also provided the opportunity to explore the impact of component changes, EMS software developments and to check that OBD software routines were correctly triggered for fault analysis.

The study was carried out in two stages, using DOE methods to define a test matrix and evaluate the results. The first stage provided a broad screening of the influence of 16 factors on the target attributes. The factors were chosen by feature teams working on the vehicle development. The factors are listed in Table 1, together with upper and lower settings of variation from the norm. The ranges of values covered by the level settings were taken directly from specifications of component tolerances for most of the sensors and actuators. In others cases, settings were based on historical data on tolerance control in production, service

C606/013/2002 © IMechE 2002

returns, and consensual judgements of worse-case possibilities. The DOE test matrix used was designed by D Grove (as Consultant to Ford Motor Company). This is a 3 level, 17 factor, fractional factorial matrix of resolution V. The lowest level of confounding is two factor interactions with three factor interactions. This allows the main effects, two factor interactions and interactions between these to be resolved. The evaluation of the matrix entailed 291 simulation runs. Each run was a simulation of vehicle operation over a cold-started NEDC test. Results for the cumulative fuel used and tailpipe emissions for each of the runs are illustrated in Figures 4(a) - (d). These give an immediate impression of the range of variations produced. In the next step, a multiple linear regression analysis was performed to arrive at a regression model. This was carried out using the Matlab Statistics Toolbox. The predictor equation reproduced the responses to an accuracy of within $\pm 3\%$. A summary of responses to lower and upper level settings is given in Table 2.

The assessment of the results was based on information provided by Effects Plots, Normal and Half Normal Plots or Daniel Plots. These highlighted the particular significance of six factors which were re-examined in the second stage of the study. At this point, the Nu-Sim model was revised to incorporate EMS software revisions and additional information on engine performance characteristics. Because of the changes, results from stage 1 and 2 were not directly comparable though none of the changes influenced the ranking of importance of the factors. The main reason for repeating the assessment of the six factors alone was to use a three level, orthogonal matrix providing greater accuracy and resolution of the effects and interactions between these. The matrix requires 45 simulation runs. Again, each run was a simulation of a cold-started NEDC test. The same process of analysis was then carried through to generate the regression model, and Effects, Normal and Half Normal Plots.

The six factors highlighted were manifold pressure sensor, EGR valve input characteristics, spark timing error, compression ratio, cylinder volumetric efficiency and EGR valve mass flow rate characteristics. The relative importance of these varies with the response considered. Thus effects plots showed that spark timing offsets of $5°$ CA from target calibration settings produced the most significant change in fuel consumption (Figure 5(a)) whilst each of four other factors had greater effect on CO feedgas emissions (Figure 5(b)). The corresponding plot for HC feedgas emissions, given in Figure 5(c), shows similar trends but lower percentage changes because of the higher baseline. The effects plot for NO_x, in Figure 5(d), shows the trends are generally inverted. This is principally a reflection of dependencies on air/fuel ratio. A summary of responses is given in Table 3. The tailpipe emissions levels are particularly prone to factor changes which cause a deterioration in mixture air/fuel ratio, and CO emissions most of all because of its strong non-linear variation with air/fuel ratio around stoichiometric mixture settings.

The Normal Plots were particularly useful in identifying factor interactions. None were ranked higher than third or fourth in importance but the effects of interactions were significant, as illustrated in Figure 6(a), (b) for CO feedgas emissions. The interaction of manifold pressure sensor and cylinder volumetric efficiency most obviously influences strategy computations of air induced, and hence mixture ratio control during transient conditions particularly. The interaction of EGR valve input, defining position, and mass flow rate for a given position, arises through the influence on EGR flow rate on charge dilution and, again, mixture air/fuel ratio. The relative importance of both interactions was much reduced in the plots for NO_x feedgas emissions, and only the second interaction was

C606/013/2002 © IMechE 2002

significant in the plots for HC feedgas emissions. The results for HC are shown in Figures 6(c) and 6(d).

4. DISCUSSION

The analysis of the DOE results produced over the two stages of the study generated a wealth of information. The objectives of the study were achieved comfortably. The approach taken proved viable to carry through. The results provided quantitative information on which an interpretation of robustness could be made. One of the major advantages provided by NuSim was repeatability and access to detail. One of the major difficulties was, and still is, how to best exploit this. In the current study, the main effects resolved in the DOE analysis were subject to further stages of investigation. The aim was to better understand the signal paths connecting factors and responses, and to identify the conditions (idle, cruise, transients, etc) when the most significant changes in fuel consumption and emissions took place. The investigations were also extended to cover off-cycle conditions simulating more severe dynamic operation conditions, and a wide range of ambient temperatures (-18°C to 40°C).

ACKNOWLEDGEMENTS

This work was carried out with the financial support and collaboration of Ford Motor Company. The robustness study involved a number of co-workers and collaborators, most importantly D Hickman and P Richardson at Nottingham and A Scarisbrick of Ford Motor Company.

REFERENCES

1. Moskwa J J, Munns S A, Rubin Z J, 'The Development of Powertrain System Modelling Methodologies: Philosophy and Implementation', SAE Paper 971089, 1989.
2. Taylor C F, 'The Internal Combustion Engine in Theory and Practice', Vol 1, 2nd Ed., MIT Press, 1985.
3. Hires S D, Overington M T, 'Transient Mixture Strength Excursions - An Investigation of Their Census and the Development of a Constant Mixture Strength Fuel Strategy', SAE Paper 810495, 1981.
4. Shayler P J, Chick J P, Darnton N J, Eade D, 'Generic Functions for Fuel Consumption and Engine-Out Emissions of HC, CO and NO_x of Spark Ignition Engines', Proc of IMechE Part D, J of Automobile Engineering, Vol 213, pp 365-378, 199 9.
5. Shayler P J, Dow P I, Hayden D J, Horn G, 'Using Neural Networks in the Characterisation and Manipulation of Engine Data', in 'Statistics for Engine Optimisation', Ed S P Edwards, D M Grove, H P Wynn; Pub. Professional Engineering Publishing, ISBN 1-8605-8201-X, pp 145-162, 2000.
6. Shayler P J, Belton C, 'In Cylinder Fuel Behaviour and Exhaust Emissions During the Cold Operation of a Spark Ignition Engine', Proc of IMechE Part D, J of Automobile Engineering, Vol 213, pp 161-174, 1999.

7. Shayler P J, Davies M T, Durrant A, Keogh P, Scarisbrick A, 'Interpreting UEGO Sensor Measurements of Mixture Air/fuel Ratio', ATA Paper 97A2IV35, Pub in Proceedings of Sixth International EAEC Congress - Conference IV Advanced Automotive Electronics, pp 1075-1088, 2nd-4th July 1997, Cernobbio, Italy, 1997.

8. Shayler P J, Hayden D, Ma T, 'Exhaust System Heat Transfer and Catalytic Convertor Performance', SAE Paper 1999-01-0453, SAE Congress Detroit, 1st-4th March 1999. Also SAE Transactions, Journal of Fuels and Lubricants, Vol. 108, 1999.

9. Shayler P J, Chick J P, Hayden D, Yuen R, Ma T, 'Progress on Modelling Engine Thermal Behaviour for VTMS Applications', SAE Transactions, Journal of Engines, Section 3, Vol 106, pp 2008-2019, 1997.

10. Lavoie G A, Imai Y A, Johnson P J, 'A Fuel Vapour Model (FVSMOD) for Evaporative Emissions System Design and Analysis', SAE Paper 982644, 1998.

Table 1: Table of Input Disturbances Applied to the 16 Factor, 291 Run Matrix, 3 Level DoE 7 Analysis. Shaded Areas are the 6 Factors Used for the 45 Run Matrix.

		Level (-1)	Level (0)	Level (+1)
1	Manifold Pressure Sensor	Lower Tolerance Limit	Nominal	Upper Tolerance Limit
2	Manifold Temperature Sensor	Lower Tolerance Limit	Nominal	Upper Tolerance Limit
3	Throttle Position Sensor	-3%	0	+3%
4	HEGO Sensor	10% Slower Response Times	Rich to Lean - 72ms Rich to Set - 500ms Lean to Rich - 65ms Lean to Set - 400ms	20% Slower Response Times
5	ECT Sensor	Upper Tolerance Limit	Nominal	Lower Tolerance Limit
6	Throttle Plate Flow Characteristics	-10% Mass Flowrate Error	0% Mass Flowrate Error	+10% Mass Flowrate Error
7	Idle Speed Bypass Valve Characteristics	Nominal	10% Leakage @ ¼ Full DC	10% Leakage @ ¾ Full DC
8	EGR Valve Input Characteristics	-4 steps Leakage	0 steps Leakage	+4 steps Leakage
9	Evap Purge Valve Characteristic	-10% Leakage @ Max DC	0	+10% Leakage @ Max DC
10	Spark Timing Error	-5° Offset	0° Offset	+5° Offset
11	Compression Ratio	10.3	10.8	11.3
12	Cylinder Volumetric Efficiency	-5% Mass Flowrate Error	0% Mass Flowrate Error	+5% Mass Flowrate Error
13	Air Intake Blockage [Coeff of Resistance]	700000	800000	900000
14	Exhaust Blockage [Coeff of Resistance]	20	30	40
15	Fuel Rail Pressure [380kPa]	Nominal –3s	Nominal	Nominal +3s
16	EGR Valve Mass Flow Rate Characteristics	-10% Mass Flow Rate Error	Nominal	+10% Mass Flow Rate Error

Table 2: Lower and Upper Error Limit Variation for Individual Design Factors 16 Factor 3 Level DoE Sensitivity Study.

		Fuel Consumption	Feedgas CO Emissions	Feedgas HC Emissions	Feedgas NOx Emissions	Tailpipe CO Emissions	Tailpipe HC Emissions	Tailpipe NOx Emissions
Manifold Pressure Sensor	Lower	-0.3%	-3.7%	-1.9%	1.9%	-15.3%	-6.6%	-6.8%
	Upper	0.2%	17.3%	2.5%	-4.4%	62.8%	19.7%	6.2%
Manifold Temperature Sensor	Lower	0.4%	-0.6%	0.4%	0.6%	-2.5%	-0.4%	0.1%
	Upper	0.4%	-1.4%	0.2%	0.6%	-7.2%	-1.4%	-0.2%
Throttle Position Sensor	Lower	-0.1%	-0.4%	-0.2%	-1.2%	-1.8%	-0.6%	-3.1%
	Upper	-0.2%	0.2%	-0.1%	1.0%	-0.3%	-0.3%	4.1%
HEGO Sensor	Lower	0.3%	1.2%	0.3%	-0.2%	2.4%	1.0%	1.0%
	Upper	0.4%	2.7%	0.6%	0.3%	8.1%	2.9%	3.6%
ECT Sensor	Lower	0.5%	-0.3%	0.5%	0.2%	-2.0%	-0.7%	1.3%
	Upper	0.4%	-0.5%	0.3%	0.2%	-4.1%	-0.1%	-0.1%
Throttle Flow Characteristics	Lower	-0.2%	-0.1%	0.0%	2.1%	1.1%	0.2%	4.9%
	Upper	0.0%	0.1%	0.1%	-1.7%	3.5%	3.5%	-7.8%
ISBV Flow Characteristics	Lower	-0.3%	-0.8%	-0.3%	0.5%	-0.5%	-0.5%	4.3%
	Upper	0.8%	-0.5%	0.5%	-0.1%	-4.3%	-1.2%	-2.3%
EGR Valve Characteristics	Lower	-0.3%	-5.0%	-1.5%	13.0%	-15.3%	-5.7%	6.2%
	Upper	0.6%	6.2%	2.4%	-9.9%	22.6%	12.6%	-0.6%
EVAP Valve Characteristics	Lower	-0.1%	0.0%	0.1%	0.0%	1.3%	1.8%	0.4%
	Upper	-0.3%	0.6%	-0.5%	-0.7%	1.6%	0.8%	-0.1%
Spark Timing Error	Lower	2.9%	4.6%	-1.7%	-11.3%	11.1%	1.0%	-22.7%
	Upper	-2.9%	-1.9%	2.3%	8.2%	0.0%	7.2%	21.8%
Compression Ratio	Lower	-0.3%	1.5%	0.6%	-4.2%	6.5%	2.7%	-4.6%
	Upper	-0.3%	-1.7%	-1.0%	4.0%	-7.8%	-3.7%	8.0%
Cylinder Volumetric Efficiency	Lower	0.0%	15.5%	2.5%	-5.2%	56.0%	21.5%	4.3%
	Upper	0.1%	-7.0%	-1.9%	3.8%	-26.8%	-11.1%	-7.0%
Air Intake Blockage	Lower	0.4%	-0.3%	0.4%	0.3%	-3.4%	-1.0%	0.1%
	Upper	0.5%	0.1%	0.4%	0.1%	-3.3%	-0.6%	1.0%
Exhaust Blockage	Lower	-0.1%	-1.2%	-0.2%	0.7%	-3.1%	-0.8%	-0.1%
	Upper	-0.3%	-0.4%	-0.2%	-0.1%	-1.2%	-0.8%	1.2%
Fuel Rail Pressure	Lower	-0.2%	-2.7%	-0.6%	0.6%	-10.1%	-2.9%	0.8%
	Upper	-0.1%	2.2%	0.3%	-0.4%	9.1%	2.5%	-2.3%
EGR Flow Characteristics	Lower	-0.4%	-6.8%	-1.6%	5.5%	-23.4%	-8.8%	-9.2%
	Upper	0.0%	9.2%	2.0%	-4.1%	38.6%	11.0%	16.8%

Table 3: Lower and Upper Error Limit Variation for Individual Design Factors 6 Factor 3 Level DoE Sensitivity Study.

		Fuel Consumption	Feedgas CO Emissions	Feedgas HC Emissions	Feedgas NOx Emissions	Tailpipe CO Emissions	Tailpipe HC Emissions	Tailpipe NOx Emissions
Manifold Pressure Sensor	Lower	-0.1%	-6.9%	-1.9%	4.3%	-6.7%	-4.5%	-11.4%
	Upper	0.0%	15.9%	3.9%	-3.9%	47.4%	15.6%	-8.7%
EGR Valve Characteristics	Lower	-0.2%	-8.6%	-2.8%	11.4%	-19.2%	-8.1%	10.3%
	Upper	0.0%	8.1%	2.0%	-11.0%	22.6%	12.9%	-7.0%
Spark Timing Error	Lower	2.7%	3.6%	-0.5%	-9.0%	8.1%	-0.1%	-4.7%
	Upper	-2.8%	1.2%	2.6%	7.6%	-3.4%	8.2%	25.6%
Compression Ratio	Lower	0.0%	2.2%	0.5%	-5.4%	3.3%	1.2%	-1.3%
	Upper	0.0%	-2.6%	-1.4%	3.3%	-7.7%	-4.1%	13.2%
Cylinder Volumetric Efficiency	Lower	-0.3%	16.8%	3.7%	-3.8%	47.4%	20.2%	1.7%
	Upper	0.2%	-10.0%	-2.1%	5.4%	-35.2%	-14.8%	-19.1%
EGR Flow Characteristics	Lower	0.0%	-9.2%	-2.4%	5.0%	-30.3%	-12.5%	-6.0%
	Upper	-0.3%	15.5%	2.4%	-3.9%	47.6%	18.3%	21.4%

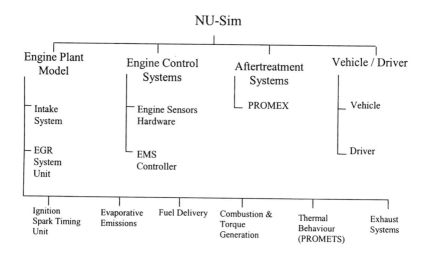

Figure 1: An Organisation Chart of the Main Sub-Systems and Top Three Levels of Hierarchy in NU-Sim.

C606/013/2002 © IMechE 2002

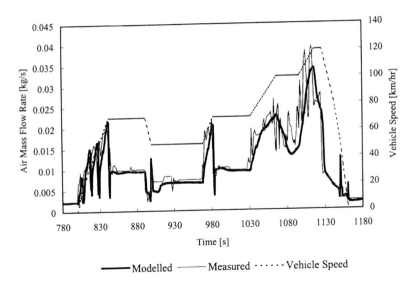

Figure 2: Induced Air Charge Mass Flow Rate, Predicted and Measured Variation Over the Final Part of the NEDC.

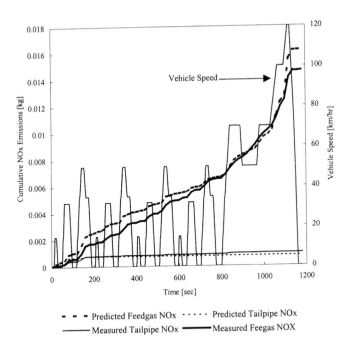

Figure 3: Cumulative NOx Feedgas and Tailpipe Emissions. Predictions and Measured Variation Over the NEDC Test.

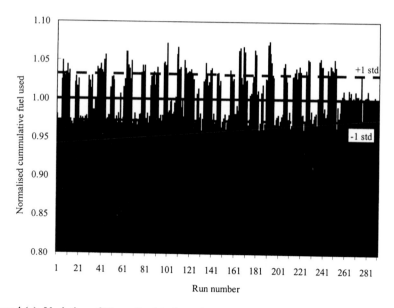

Figure 4 (a): Variation of Normalised Indicated Fuel Consumption to Baseline Value for the Range of the Experimental Matrix.

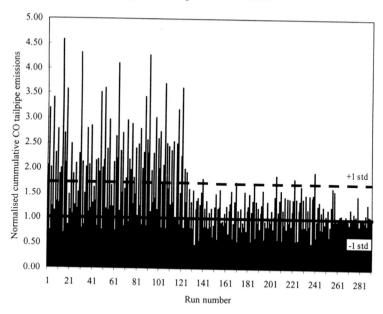

Figure 4 (b): Variation of Normalised Cumulative Tailpipe CO Emissions, to Baseline Value, for the Range of the Experimental Matrix.

C606/013/2002 © IMechE 2002

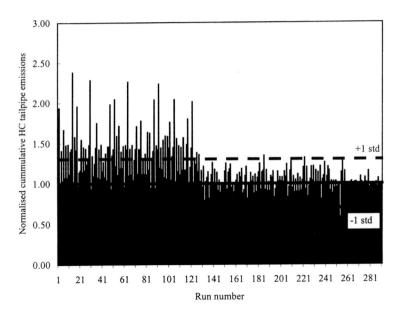

Figure 4 (c): Variation of Normalised Cumulative Tailpipe HC Emissions, to Baseline Value, for the Range of the Experimental Matrix.

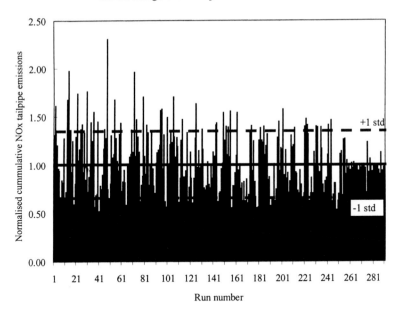

Figure 4 (d): Variation of Normalised Cumulative Tailpipe NOx Emissions, to Baseline Value, for the Range of the Experimental Matrix.

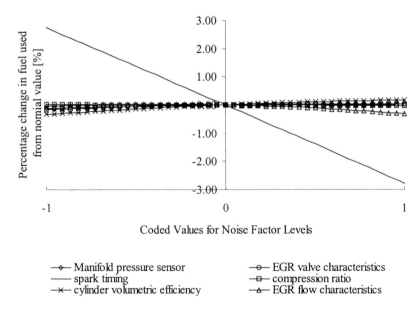

Figure 5 (a): Effect Plot Showing the Effect of Positive and Negative Errors in Spark Timing on Fuel Consumption Along Side the Other 5 Input Disturbances. Results are for a NEDC at 20°C Start Temperature.

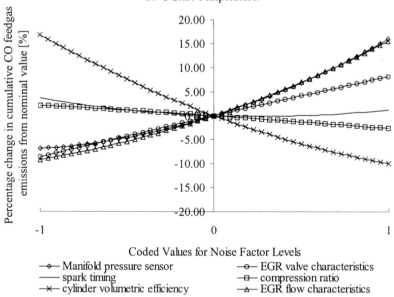

Figure 5 (b): Effect Plot Showing the Effect of Positive and Negative Factor Errors on Cumulative CO Feedgas Emissions. Results are for a NEDC at 20°C Start Temperature.

C606/013/2002 © IMechE 2002

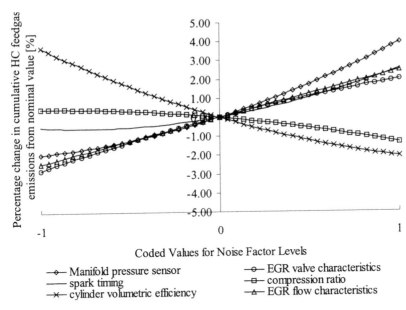

Figure 5 (c): Effect Plot Showing the Effect of Positive and Negative Factor Errors on Cumulative HC Feedgas Emissions. Results are for a NEDC at 20°C Start Temperature.

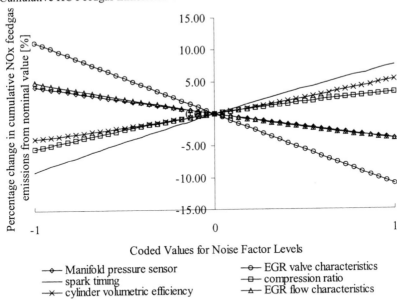

Figure 5 (d): Effect Plot Showing the Effect of Positive and Negative Factor Errors on Cumulative NOx Feedgas Emissions. Results are for a NEDC at 20°C Start Temperature.

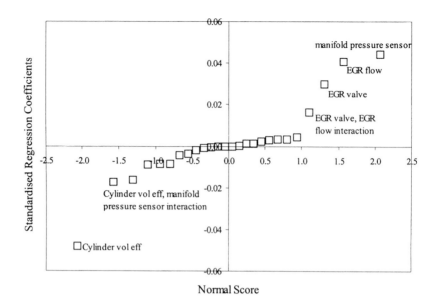

Figure 6 (a): Full Normal Plot for CO Feedgas Emissions.

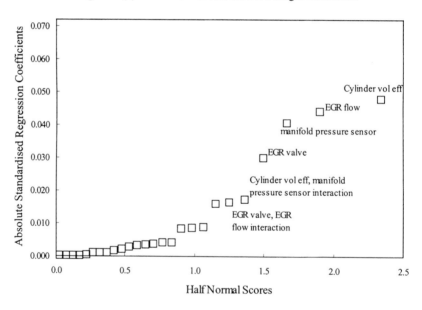

Figure 6 (b): Half Normal Plot for CO Feedgas Emissions.

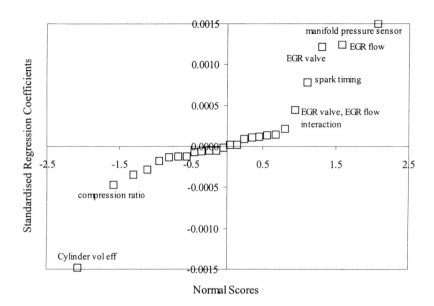

Figure 6 (c): Full Normal Plot for HC Feedgas Emissions.

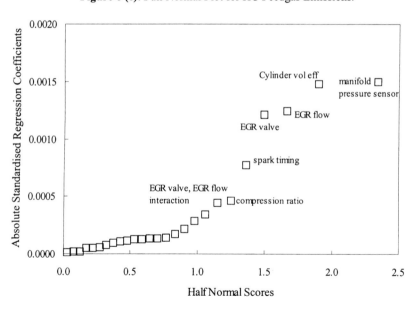

Figure 6 (d): Half Normal Plot for HC Feedgas Emissions.

C606/015/2002

The application of multivariate correlation techniques to driveability analysis

S G PICKERING, C J BRACE, and **N D VAUGHAN**
Department of Mechanical Engineering, University of Bath, UK

SYNOPSIS:

The aim of the work described in this paper is to investigate the application of multivariate regression techniques to driveability analysis. Vehicle driveability is difficult to quantify in an objective sense as it is based on a driver's subjective rating of a vehicle. The ability to predict the subjective driveability rating for a vehicle using only objective parameters such as acceleration, jerk and throttle demand will allow manufacturers to calibrate vehicle powertrains far faster than is presently possible. It will also allow greater scope for vehicle characterisation and allow simultaneous emissions, economy and driveability constraints to be met more easily. This paper presents a methodology for producing correlations between subjective ratings and objective driveability data. It describes the preliminary code used to automate the process and the results of using simple regression techniques to analyse longitudinal driveability.

1. INTRODUCTION

A vehicle's driveability rating is based on the driver's subjective reaction to the vehicle's objective behaviour. The ability to link together the subjective driveability rating with objective data is an area of automotive engineering which is attracting growing interest wherever the driver and vehicle require a close interaction. This ability will facilitate a number of tasks that are currently time consuming and require iteration due to the subjective nature of their assessment.

One of the most important of these will be the ability for engineers to perform accurate driveability calibration on vehicle powertrains whilst still on the test bed or even at the simulation stage, thus reducing calibration time and redesign costs. Other benefits include the ability to characterise and benchmark the driveability performance of a wide range of vehicles, which might for example allow a calibration engineer to copy a vehicle with particularly good driveability.

Usually vehicle driveability rating is performed by groups of test drivers consisting of a cross-section of the population for whom a vehicle is designed. This might include both experienced and novice test drivers, as well as both men and women in different age groups. These drivers perform a pre-determined set of tests with the vehicle that are designed to highlight certain aspects of driveability. After each of the tests the drivers fill in a questionnaire which asks them to rate their opinion of aspects of the vehicle's performance. Each aspect is given a rating often on a scale from 0 to 10 (1, 2) which denotes how good or bad a certain aspect seemed to the driver. Inevitably the human factor in determining a subjective rating is prone to scatter even for one individual. Additionally differences in the perception of the different groups can make the task of developing correlation methods very difficult. In a research context fewer drivers with similar backgrounds and experience may be used to collect data in order to reduce this scatter. This can facilitate the establishment of an effective methodology by reducing the amount of data required to represent the selected population. The subjective rating data provided by the test drivers is used with objective data simultaneously recorded during the test. The goal in the work described here is to find correlations between the subjective ratings and the objective data.

One difficulty is that the format of these two data sets is very different. The objective data will be a time based recording of a number of channels describing the vehicle response which may correspond to a single-figure driver rating. Hence for correlation analysis, these data must be processed to produce metrics which characterise the objective performance in more succinct form. For example the acceleration response of the vehicle could potentially be represented by a single figure for peak acceleration.

The selection and calculation of driveability metrics is at the heart of the correlation analysis. Dorey and Martin (3) mentioned their use of the following metrics in the analysis of longitudinal acceleration:

- Acceleration response overshoot
- Rise rate
- Damping during decay of oscillations

Wicke (4) highlighted the following metrics as being important in his correlations:

- Acceleration delay time
- Initial and maximum acceleration
- Initial and maximum jerk (rate of change of acceleration)

In this analysis it is assumed that the objective metrics are independent variables because each driver must perform a sequence of pre-defined tests, and that their driveability ratings are therefore dependent variables. Simple correlations (i.e. between two variables) produce easily understood results, but they do not take into account the fact that many different driveability metrics may all affect a given subjective rating. To be able to analyse the effect of multiple metrics on a single subjective rating multivariate correlations must be used. These techniques correlate a single driveability rating with a number of metrics, and thereby allow the relative effects of these different metrics to be determined.

C606/015/2002 © IMechE 2002

In addition to the establishment of the metrics the same workers have investigated potential correlation methods. Wicke from the University of Bath, showed that applying simple correlations to longitudinal driveability analysis (4) produces a number of definite correlations. Dorey et al from Ricardo (2, 3) have also been working in the area of driveability analysis. Like the work carried out at AVL, the work discussed by Dorey et al covers a broader region of driveability than that investigated in this paper, however they do not mention the application of multivariate techniques.

List and Schoeggl from AVL (1) mention their use of some multi-dimensional techniques in their driveability analysis. The multi-dimensional techniques were applied to a broad range of vehicle performance metrics in concert with a neural net to try to simulate a human's subjective reaction. However it appears that the main part of the work was concerning the use of neural nets rather than regression equations.

The current paper investigates a method by which different objective metrics can be ranked in order of their correlation with subjective ratings. By finding those metrics which are most closely correlated with subjective ratings, calibration engineers will be able to concentrate on optimising only a limited number of objective parameters. Using these highly correlated metrics will also facilitate both multivariate regression and neural net analyses by reducing the number of objective metrics which need to be included or tried in such analyses.

These correlations are of course a simplification of the true relationships between the metrics and ratings, as they only indicate the relationship between one rating and one metric, however this work shows which variables are likely to be important for use in a multivariate regression analysis. The determination of highly correlated metrics also allows calibration engineers to make vehicle data acquisition more focused by reducing the type and amount of data which needs to be recorded in each test.

2. CORRELATION METHODS

There are two main multivariate approaches that can be applied to analysing the correlations between the subjective rating and the metrics representing the objective data.

The first covers iterative methods that may be used to determine either a correlation equation or a representative system model. This category includes techniques such as genetic algorithms and neural networks. The second uses statistical regression methods to establish a system equation and this includes least squares methods with the use of correlation indices to rate the results.

Both methods have advantages and disadvantages. Iterative methods, by their nature, require time and inclusive data sets to produce a solution. In the case of neural nets, large amounts of training data and time are required so that the internal structure of the net can be established. However neural nets can simulate very complicated equations due to their internal flexibility. Regression techniques require less training time and data, however they are not as flexible as neural nets, and require external input in the form of selecting the appropriate type of regression equation. However this relative simplicity also means that once a regression equation has been produced, it is far easier to determine the effects that the different inputs have on the output. This simplicity will allow a calibration engineer to see more clearly which calibration aspects affect powertrain performance ratings and to what extent. This will also allow informed decisions to be made on what trade-offs can be made for emissions and economy and their effect on driveability.

Although using neural nets remains a more flexible technique for simulating complete vehicle driveability, assuming the availability of a comprehensive data set, the use of statistical regression equations may prove useful in assessing specific areas of driveability such as longitudinal driveability. Both List and Schoeggl (1) and Dorey et al (2,3) have included many aspects in their driveability analysis including engine start and warm-up behaviour, the effect of gear changes and the application of accessory loads. By excluding the effects of these variables, and concentrating solely on those variables immediately affecting longitudinal driveability, the statistical regression approach will be able to produce good correlations which will be easier to analyse than those produced by a neural net with many inputs.

3. IMPLEMENTATION

The analysis of large amounts of data requires that it be automated as much as possible, so that the process can be performed quickly, reliably, and accurately. To this end, the current project has focused on automating both the processing of raw data files (from vehicle data acquisition for example), and the analysis and correlation of the data that is recorded in these files together with subjective ratings so that the user is simply presented with the list of correlation results.

As a first stage, the large data files containing the sampled objective data are processed. The calibration of the objective data is first checked for both acceleration and pedal position offsets and if necessary automatically adjusted without user intervention. Zero-phase digital filtering is applied to the acceleration data to provide smoothing and facilitate the extraction of driveability metrics. The user is notified of any errors found and changes made to the data.

3.1 Extraction of driveability metrics

Various driveability metrics are automatically calculated using the data within these files. The choice and calculation of the metrics can be easily altered or added to, due to the modular nature of the code. The code automatically processes the results from a complete set of test runs (i.e. all the tests performed by one driver) and output a separate data file for each test containing the calculated metrics and subjective ratings.

This procedure has been developed to be non-specific. This means that data from different data acquisition systems containing different variable channels, with different sampling rates and calibrations, can still be used with the correlation code.

As an example demonstrating the procedure a selection of generated objective metrics for an acceleration demand or "tip-in" manoeuvre is listed below:

- Maximum vehicle acceleration
- Initial acceleration
- Rate of change of acceleration

C606/015/2002 © IMechE 2002

- Delay time (between pedal movement and start of acceleration)
- Maximum vehicle velocity.

Figure 1 shows a graph of vehicle acceleration and pedal position against time. Time 1 indicates the start of the pedal movement, Time 2 indicates the start of vehicle acceleration, and Time 3 coincidentally shows both the end of the initial acceleration period, and the maximum acceleration point. These times and magnitudes are automatically determined by the code, and are used to generate the driveability metrics. It should be noted that this list is not exhaustive and that other times and magnitudes are also calculated and used to generate these and other metrics. The initial acceleration (Time 1) and pedal movement times (Time 2) are calculated by setting boundaries around the initial values within which small oscillations are ignored. The starting time is assumed to be the point at which the data moves outside this range. The maximum acceleration (Time 3) is found very simply using a standard MATLAB function. The end of the initial acceleration phase (which is also indicated on Figure 1 by Time 3) is calculated by iterating through the acceleration data and determining when the acceleration gradient changes significantly. A more accurate technique being investigated uses the magnitude and sign of the gradient of the data to determine the starting points.

The driveability data used in this research was originally recorded to determine those aspects of AT performance which drivers liked, so that these could be applied to an experimental CVT (4). The recorded data include both gearshifts and kick-downs, however as they were not of interest at the time, no effort was made to interpret this information.

The present work is also focused on longitudinal aspects of driveability rather than gearshift aspects and therefore these are not interpreted. The current approach could be used for the analysis of data from MT vehicles, however it would require that the testing be carried out either in a single gear or by experienced drivers whose gear shifting technique showed little variability.

Shift quality is a very important aspect of driveability for both ATs and the newer Automatic Manual Transmissions (AMTs) and, although it has not been addressed in the present work, it would be desirable to expand the research in this area.

3.2 Regression Analysis

A correlation code has initially been developed to produce single variable linear and curvilinear correlations. Simple sub-grouping of the metrics files is also automated, meaning that the test results for a single vehicle, driver, vehicle initial velocity, pedal position or some combination of these can quickly be analysed in isolation.

The correlation methods employed in the current code are all based on least squares fitting of various regression equations (see Table 1) to the recorded data points. The least squares technique determines the best parameters for any given equation, so that the mean square-root of error between the calculated points and the actual data is minimised.

Equation Type	General Form
Linear	$Y = a + bX$
Parabolic	$Y = a + bX + cX^2$
Cubic	$Y = a + bX + cX^2 + dX^3$
Log (1)	$Log (Y) = a + bX$
Log (2)	$Log (Y) = a + b.\ Log (X)$
Log (3)	$Y = a + b.\ Log (X)$
Hyperbolic	$Y = \dfrac{1}{a + bX}$

Table 1 – Regression Equations

3.3 Index of correlation

The index of correlation (5) is used as a measure of how closely a given curvilinear equation correlates the value of the subjective rating to the objective metric. This index of correlation is similar to Pearson's product-moment coefficient of correlation (6) but for curvilinear rather than linear regression equations. The index is a number between 0 and 1, where 1 shows an exact correlation and 0 shows no correlation. The index of correlation is calculated using the following equation:

$$i = \frac{S_{y'}}{S_y}$$

Equation 1

[Where $S_{y'}$ is the standard deviation of the subjective rating data points predicted by the regression equation from the actual objective metric data, and S_y is the standard deviation of the actual subjective rating data points.]

These techniques will be extended for use in the multivariate case to determine both the total correlation and partial correlations within a set of objective and subjective variables.

4. EXPERIMENTAL METHOD

This correlation technique has been tested using data acquired during a previous project at the University of Bath (7). The results presented below show that the suggested method produces correlations that are similar to those reported by Wicke et al in their papers (4, 8) analysing a similar set of driveability data.

4.1 Data Acquisition Method

Data were collected from a number of vehicles which are described in Table 2.

C606/015/2002 © IMechE 2002

Vehicle	Engine Type	Transmission Type	No. of Test Drivers
Car A	2.5l Petrol	AT (5 speed)	18
Car B	2.0l Petrol	Experimental CVT	13
Car C	2.0l Petrol	AT (4 speed)	14

Table 2 – Test Vehicles

A group of test drivers performed a series of tests, which are described in Table 3, with each of the test vehicles. Each test had a specific vehicle starting velocity, which the driver had to reach before the test was started. Once a steady speed had been reached, data acquisition was started and the driver moved the accelerator pedal quickly to a set position depending on the test type. The accelerator pedal was kept at the specified level for the remainder of the test to allow transients to die away. Objective data were recorded for 8 seconds during the test, and immediately after each test the driver was asked some subjective questions relating to the vehicle's performance.

Test Type Number	Test Description	
	Initial vehicle velocity (km/h)	Desired accelerator pedal position (%)
1	0	25
2	0	50
3	0	75
4	0	100
5	12	25
6	12	50
7	12	75
8	12	100
9	40	75
10	40	100
11	60	75
12	60	100

Table 3 – Test Descriptions

After each test, the drivers were asked to give the vehicle a rating between 0 (very bad) and 10 (perfect) for each of the categories listed below:

- General Driveability
- Jerk
- Overall Acceleration
- Engine Delay
- Vehicle Delay
- Smoothness

Often the word *jerk* is used to describe a negative aspect of performance such as driveline shunt or poorly timed clutch engagement which cause oscillatory movements in the vehicle. However, in this work, *jerk* is the name given to the initial rate of change of acceleration. This is considered to be a positive aspect of performance giving an indication of the speed with which the acceleration builds.

5. RESULTS AND DISCUSSION

In his papers (4,8), Wicke reported a number of different correlations between the subjective and objective data which had been recorded. He found that the subjective driveability rating was correlated with the following objective parameters:

- Delay time
- Initial acceleration
- Jerk

Unsurprisingly, Wicke found that subjective acceleration related ratings (acceleration and jerk) were positively correlated with objective acceleration parameters. Figure 2 shows the subjective jerk rating plotted against maximum acceleration for all vehicles and tests starting at 0km/h. This shows a degree of correlation as expected. Also included on this graph are various regression lines which were calculated by the correlation code. The solid line is that which produced the highest index of correlation as can be seen from Table 4 below. There is a degree of scatter in the data as noted in the introduction, which reduces the indices of correlation, analysing the data for the vehicles individually may reduce this scatter, otherwise it may be due to the influence of other objective metrics which would be taken into account with a multivariate approach.

Linear	Cubic	Parabolic	Log1	Log2	Log3	Hyperbolic
0.319972	0.321342	0.321179	0.301608	0.299402	0.310138	0.269858

Table 4 – Correlation Coefficients

Wicke also found that subjective driveability and initial acceleration ratings were related to both objective acceleration, jerk and delay time. It was indicated that the subjective delay time rating is also influenced by objective acceleration. Figure 3 shows a graph of subjective vehicle delay rating plotted against the objective maximum acceleration recorded during the test. The solid line is the cubic regression equation which produced the best index of correlation of 0.618633. This graph clearly shows a correlation between the delay time rating and the maximum vehicle acceleration.

These results are generally in line with expectations, for example subjective driveability ratings are highly correlated with various objective acceleration and jerk parameters. The rating for smoothness is also highly correlated with acceleration and jerk as shown in Figure 4.

    C606/015/2002 © IMechE 2002

5.1 Problems

It should be noted that development is still in progress on the code and methods presented in this paper. There are a number of shortcomings with the current implementation which will be addressed in future work to enable these methods to have real analytical value.

At present there is no indication to the user of the statistical significance of a given index of correlation. This is problematic because for a small number of points, and a sufficiently high order fitting equation, the curve which is produced can pass through all of the data points, producing a perfect correlation. The fitted equation is obviously then only valid for this small set and does not mean that there would be a strong correlation if more data points were available. Therefore, depending on their statistical significance, the correlations need to be either adjusted, excluded or clearly marked for the user.

Another related problem with the current method is that for arrangements of data points which form groups (i.e. with large gaps between the groups), the regression line can fluctuate in the y dimension (the subjective rating). This can be seen in Figure 5. In this case it is possible that the two groups of data points should be analysed separately, which would occur if a multivariate approach were used, or some more data should be acquired specifically to fill the gap between the two groupings and produce a more accurate regression equation. In extreme cases this grouping can cause the regression equation to stray out of the allowable subjective limits (the range 0 to 10) producing invalid subjective ratings for certain values of the objective metric. This is obviously not acceptable and the problem could be solved by implementing constraints in the fitting algorithm or by testing the data for clustering before fitting a regression equation.

It can also be seen from Figure 6 that outlying data points are able to significantly skew the regression equation that is used to calculate the correlations. This is due to the least sum of squares fitting method which is used in the current code. This technique was selected because it is simple to implement using MATLAB, however there are more robust techniques such as the least median of squares method (9).

6. CONCLUSIONS

The code already produces meaningful correlations, and will improve with a more robust regression equation fitting method which will reduce the impact of scatter and outlying data points.

The code enables large amounts of data to be processed quickly, and without the user needing to understand the underlying processes and programming.

The modular structure of the code also enables the user to expand the regression equation facilities to suit their own preference and to more easily specify which subjective and objective parameters should be correlated against one another.

The current code is a good starting point for the development of both a multivariate analysis system and a comparative neural net system.

7. FUTURE DIRECTION

The current parameter generation code is already very useful in processing large volumes of data requiring little or no user intervention, however there is always scope for improving the methods by which the metrics are generated. In particular, the delay time metric is difficult to determine accurately, especially while the vehicle is moving and may suffer from minor accelerations and decelerations after data acquisition has started but before the driver has moved the accelerator pedal. Other aspects of longitudinal driveability will also be considered, for example deceleration demands, overrun braking feel and traffic crawl performance will be analysed.

The correlation code, in its current incarnation produces structured, easily understandable results; both in graphic form if required, and in tabular form.

A pre-processing step will be included to identify and remove inter-relationships between the subjective rating variables and the objective metrics respectively. This may take the form of a Principal Component Analysis (PCA). This will increase processing speed and reduce the complexity of analysing the results.

The correlation code will be developed to perform multivariate correlations. This will show both the effect which different objective factors have on one another, and the way in which the individual objective factors affect one another and thereby the subjective rating. A neural net will also be developed for comparison with the multivariate system using the current correlation code as a pre-processing stage.

REFERENCES

1. List H., Schoeggl P. (1998). *Objective Evaluation of Vehicle Driveability. SAE 980204.*

2. Dorey R.E., Holmes C.B. (1999). *Vehicle Driveability – Its Characterisation and Measurement. SAE 1999-01-0949.*

3. Dorey R.E., Martin E.J. (2000). *Vehicle Driveability The Development of an Objective Methodology. SAE 2000-01-1326.*

4. Wicke V., Brace C.J., Deacon M., Vaughan N.D. (1999). *Preliminary Results from Driveability Investigations of Vehicles with Continuously Variable Transmissions. CVT'99 - International Congress on Continuously Variable Power Transmission September 16-17, 1999, Eindhoven University of Technology.*

5. Ezekiel M., Fox K. A. (1966). *Methods of correlation and regression analysis,* p.128. 3rd ed. New York: John Wiley and Sons, Inc.

6. Hays W.L. (1988). *Statistics,* pp.555-558. 4th ed. New York: Holt, Rinehart and Winston, Inc.

7. Wicke V. (2001). *Integration and Control aspects of CVT vehicles.* Ph.D. thesis, University of Bath.

8. Wicke V., Brace C.J., Vaughan N.D. (2000). *The Potential for Simulation of Driveability of CVT Vehicles. SAE 2000-01-0830, SP-1522.*

9. Rousseeuw P. J., Leroy A.M. (1987). *Robust Regression and Outlier Detection.* New York: John Wiley and Sons, Inc.

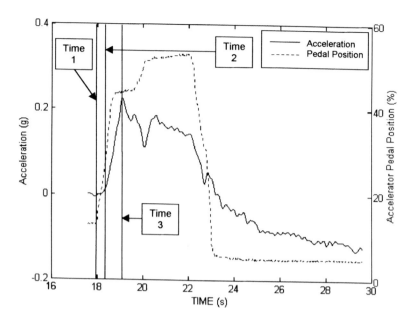

Figure 1 – Explanation of driveability metrics

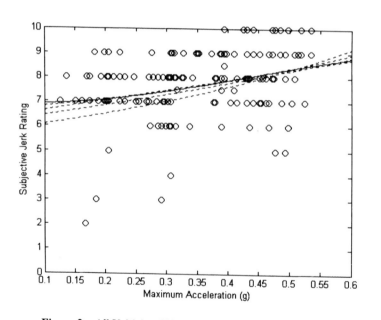

Figure 2 – All Vehicles, 0Km/h initial test velocity (tests 1-4)

C606/015/2002 © IMechE 2002

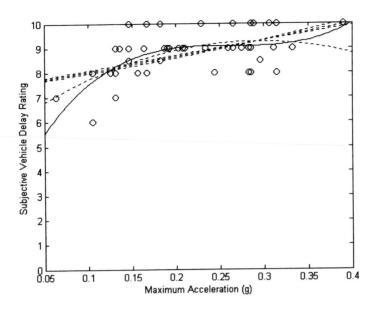

Figure 3 – Vehicle A, 12Km/h initial test velocity (tests 5-8)

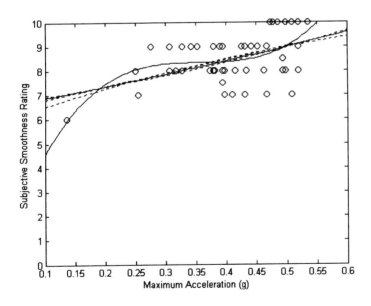

Figure 4 – Car A, 0Km/h initial test velocity (tests 1-4)

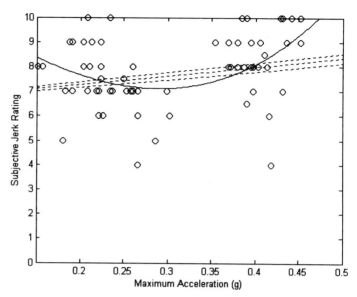

Figure 5 – All vehicles, 40Km/h initial test velocity (tests 9 & 10)

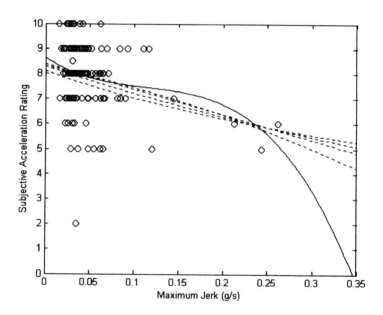

Figure 6 – All vehicles, 12Km/h initial test velocity (tests 5-8)

Vehicles, Systems
and Components

C606/024/2002

Experiencing customer vehicle comfort requirements using concept synthesis techniques

D J FOTHERGILL and **M ALLMAN-WARD**
MSX International Limited, Millbrook, UK

1   SUMMARY

Today's design process is driven by brand image coupled with customer expectations. In order for designers to assess whether they will delight or even satisfy their customers, it is necessary to gain an appreciation of the performance of their components evaluated in the true context, not represented as functions on a graph.

This paper describes an engineering process developed to enable customer requirements to drive the vehicle design. It discusses the reasons for concept modelling and making validated assessments of what our models tell us, looks at how we can assemble simulations in a simple and timely manner (using readily available data), and illustrates what we can gain by experiencing simulations.

2   PRODUCT DESIGN PROCESS – THE KEY DRIVERS

It has long been realised that Products are bought and sold by their character or "brand values", most obviously through marketing slogans, but it is only recently that these values have been truly used as the key driver of the design process.

From a manufacturer's perspective, vehicles are designed to meet the expectations of a loyal existing client base, or to attract new customers to the brand. To do this they must possess characteristics which are either unique to the brand (and hence appealing to the targeted customers), or are better than those of competitors. In order to deliver a vehicle which reinforces the corporate brand image, it is necessary to start by understanding the relationship between the brand values and measurable attributes. These form the basis of objective targets for the vehicle designers. By re-synthesising these targets into a form that can be subjectively assessed, for example by listening to the sound of the whole vehicle expressed as the sum of its targets, these targets can be validated as delivering the brand image. It is only through this process that the brand character of a vehicle is attained predominantly by design rather than by development.

3   CONCEPT SELECTION & VEHICLE DEVELOPMENT

Corporate knowledge and legal requirements are combined with the brand values to define the top level objectives for a project (Figure 1). These objectives are then translated by attribute into various vehicle targets for the engineering functions. The key attributes typically include styling, vehicle dynamics, NVH, performance and economy, safety and durability.

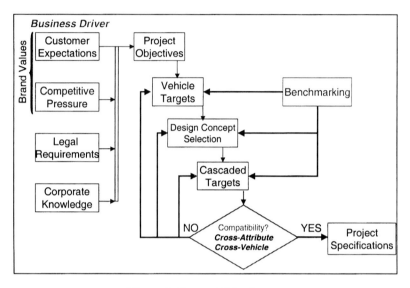

Figure 1: Concept Selection

Best candidate concepts are then progressed by cascading these targets to the point where sub-system, or even component, targets can be produced. At this stage concepts may be accepted or rejected on the basis of their ability to meet cross-attribute compatibility criteria. This does not require production of detailed component models, but simply determining whether the resultant component targets are practical. This may be assessed by benchmarking existing component designs, or by performing analysis using simple models. Once it has been ascertained that the component targets are practical, it is also necessary to assess whether their combination delivers the whole vehicle target. It should be noted that not all targets can be expressed totally objectively. For example, customers express their holistic experience of the vehicle in terms of descriptive words. In order to assess whether the subsystem targets deliver the desired brand values, some may require synthesis from analytical results to physical stimuli in order to be experienced by human assessors through subjective evaluation.

The above process will result in a complete set of unique parameters for a particular vehicle. Economic arguments could, however, lead to modifications of certain components in a vehicle to facilitate their use on another platform. The process can therefore be made more complex by the need to meet targets from more than one design, to achieve cross-vehicle compatibility. This could be as simple as adding extra bolt holes in a subframe, or as radical as adding tailor welded blanks to a pressing to let a niche design share parts from a mainstream vehicle.

This process must however be applied flexibly. Although individual parts often dominate one aspect of the vehicle's characteristics, ultimately the requirement is that a component will perform its role as a part of the whole vehicle system. For example, an engine will deliver a predefined power curve with a particular specific fuel consumption irrespective of whether it is installed in a vehicle or is mounted on a test bed. The vehicle performance and economy will however also depend upon gear ratios, wheel and tyre properties, aerodynamic drag properties, vehicle mass, cooling etc. It is therefore necessary to trade and compromise between these attributes to deliver the targets not only for Performance & Economy, but also for the other attributes as well. This can only be achieved by concept modelling early in the program.

A further advantage of concept modelling within this process is that designs which are tuneable can be selected over those which are not. If the design is tuneable, then the effect of certain changes to a component to deliver one attribute target have a limited and understood effect on other attributes. That is, the parts are designed and tuned together to control the performance of a vehicle to meet all its attribute-led targets, rather than designing each component to meet a specific vehicle target and then analysing the design to see how it works for other attributes.

To elaborate on the previous example, it may be possible to share an entire powertrain design, complete with mountings, between quite different vehicles. To make the performance and economy of each vehicle match the respective brand values it may be necessary to change gear ratios and / or engine control mapping. To tune the way each vehicle sounds, attention is likely to be required to the mount system, detailed local body design, exhaust and air intake systems. If these systems are tuneable then, for example, changing the mounts need not necessarily adversely affect driveability or secondary ride.

By implementing a target cascade, validation and attribute trading approach, the vehicle targets can be balanced at the concept stage, based upon what is desirable versus what can be practically achieved through tuning the parameters of the design. Given a proper application of this process, the required degree of tuning (and its impact on other attributes like ride or performance) would be known at the outset, rather than gained by a process of analysis and experimentation late on in the design.

4 TIMING

Currently, the time taken from the start of the concept phase through to its conclusion will typically be around six months, after which most of the vehicle attributes become practically non-tradeable. In a further five months (or less) system targets will themselves have become embedded in the design, and the component manufactures will be seeking to produce viable designs. A further three months on, the engineering approach will more or less have stabilised, and all efforts will be focused on bringing it all together and ironing out problems, modifying and rebalancing to meet production needs etc.

That is, by fourteen months into the process any opportunities for major revision will have completely disappeared, despite the fact that it will only really be at this point that sufficient complete vehicle design data will be available to make a complete analytical assessment.

5   VALIDATED TARGETS ARE THE KEY TO CONCEPT ENGINEERING

As mentioned earlier, problems arise when assembling data from which to synthesize component models, because there will be little or no detailed modelling or hardware available at the concept phase. This can result in the work not being performed, preventing the assessment of the target based model. However, application of the target cascading process in conjunction with parameterised modelling allows the shortfall in detailed information to be replaced by synthetic components, i.e. validated targets, and hence allows CAE modelling to lead instead of lag the design.

As an example of engineering with no hardware, vehicle level NVH targets are largely determined using subjective jury evaluation techniques. Simulators are used to play sounds or vibrations (or combinations of both) to representatives of the client group (typical customers and/or brand managers). The results of these subjective evaluations are compared to objective metrics of the stimuli using statistics to generate objective targets which satisfy the brand and customer expectations. These targets can be cascaded down to sub-system and even component level, adjusted to reflect concept models, traded, and re-synthesised to the whole vehicle sound at any time to verify that the sound of the concept will deliver the brand.

6   SUMMARY OF THE ATTRIBUTE DRIVEN DESIGN PROCESS

In summary, the process described above is:

1) Top down
- Define the vehicle attributes
- Convert the vehicle attributes to vehicle targets
- Cascade the vehicle targets to systems, subsystems and components

2) Bottom up
- Parameterise component / sub-system designs
- Synthesize the components (or subsystems, or systems) to meet cascaded targets
- Assemble the synthesized parts into a complete vehicle synthesis
- Make objective and subjective assessments of total vehicle performance against vehicle targets

3) Tuneable
- Adjust individual targets as necessary to achieve balance/desired vehicle result.

The attribute-led process allows the market needs to drive the design. The systems, subsystems, and components are all designed to meet cascaded targets that are consistent with achieving the desired vehicle performance. The result will be a vehicle that is specified from the top down and designed to that specification from the bottom up.

The construction of concept, rather than strictly physical, models facilitates the simulation of the effect of all of the components meeting (or not meeting) their targets. Thus, the target balance may be subjectively or objectively assessed and adjusted. This process facilitates designing for the customers, instead of trying to find customers for what has been designed.

7   CONCEPT ANALYSIS

Components in the concept model are synthesized such that they meet the cascaded targets, or can be easily tuned to do so. The first objective of the concept level vehicle model is to provide a means of assessing the balance of cascaded targets before designers begin to draft parts.

This does not require the vehicle component physical attributes to be modelled, but rather their behaviours or functions. By creating parameterised models of components (or entire systems), the desired target balance can readily be assessed. Once the parameterised component models are seen to work together to create desirable system characteristics, their attributes can be issued as validated cascaded targets.

A classic example of the top down, bottom up approach to concept design via parameterised modelling is the hydraulic engine mount. Even in a full vehicle simulation, such a component would rarely be included as a physical model with solid representations of the rubber and metal parts, etc. Typically, such a component is broken down into its underlying parts, with an independent degree of freedom attributed to the fluid mass and scalar representations of the structural and volumetric stiffness and damping (Figure 2). A parameterised model of the system is generated to represent that component which can be validated against known parts. Guidelines are taken from mount manufacturers (or from reference to a spread of known parts) to limit the parametric model to what is achievable, but the essential behaviours of the component are fully represented by the parametric model and can be tuned at will. The component model parameters can therefore be optimised to the requirements of the whole vehicle model in a simple, effective manner. Using the same model, the resulting parameters can then be translated back to physical design values, and confirmed as component targets.

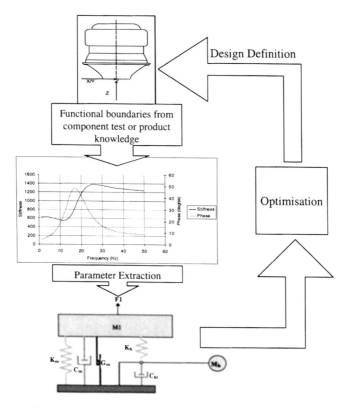

Figure 2: Component Parameterisation Example - Hydromount

The body shell can be similarly mathematically parameterised. It is possible to create a generic model, of variable mass and modal properties, which can be adjusted to approximate to the intended design. The modal parameters can then be varied to study the effect of interaction with other vehicle systems. For example, it is a common misconception that the body shell should be as stiff as possible, with consequently the highest achievable bending and torsional frequencies. In practice, this approach can yield severe vibration problems. The most logical approach is to deliberately adjust frequencies (not necessarily making them as high as possible) and control the shapes of modes of vibration of the body shell such that dynamic interaction with other vehicle systems is harmonised.

C606/024/2002 © IMechE 2002

Once the body shell fundamentals have been validated in the vehicle system model, a more physical representation can be developed, utilising a coarse combination of beams and shells, with springs at the joint locations (Figure 3). Although containing physical elements, such a model is still virtually fully parameterised, in that for example beam sections and joint stiffness are still scalar quantities that can be readily changed. This model can then be tuned until its dynamic attributes reach those values validated by the original generic model, thus giving valuable design guidance regarding sizing of structural components and the influence of the positioning of major items of mass etc (e.g. battery, seats...).

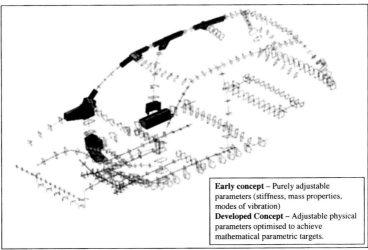

Early concept – Purely adjustable parameters (stiffness, mass properties, modes of vibration)
Developed Concept – Adjustable physical parameters optimised to achieve mathematical parametric targets.

Figure 3: Component Parameterisation Example – The Bodyshell

At the concept stage, the suspension systems may be represented by flexibly connected rigid bodies. It may be valid to ignore the influence of link deflection in the suspension dynamics given that they should be designed not to participate flexibly in the first place. Provided that the link masses and inertias are representative, this level of model may be used to understand and develop the influence of the bush stiffness values on the dynamic performance (e.g. road noise isolation). Once the bushes are coarsely tuned, the component stiffness targets may be derived in order to meet the above assumption. As the design develops, the rigid links will be replaced with models derived from physical designs for the purposes of validation.

When the suspension geometry is approximately mapped out, the steering system can be added. Studies are then performed to assess the likely degree of road wheel to steering wheel interaction via the rack and column system for alternative geometry and isolation schemes.

It is always possible to assemble models of vehicle concepts ahead of completion of the detailed design. This is essential if the models are to be used to lead and develop the design rather than merely to validate it, ensuring that the design delivers the required performance attributes. It also enables those involved in design and brand management to experience physically the predicted model performance using simulation techniques. Informed subjective judgements can then be made regarding the validity of targets, and the cost / benefit of the investment required to meet them.

Just as a detailed vehicle model is used to assess aspects of the performance of an existing design, and to perform troubleshooting type studies, so the concept model is used to validate and develop the cascaded target balance. Instead of predicting the characteristics of an existing design, the engineer can evaluate how the design would perform if a particular target cascade were met.

8 CONCEPT LEVEL SUBSYSTEM TARGET BALANCING STUDY

The authors have frequently used hybrid models constructed from a combination of measured and subsystem target data in order to both assess and develop the target balance.

The most typical cross attribute compatibility problems arise from the need to design multifunctional components, for example a bracket that has to support an engine mount and locate a suspension bush. For that matter, the engine mount itself has a dual role, comprising flexibility to provide isolation from engine induced vibration (combustion, out of balance etc.), stiffness and damping to control shake, and sufficient ultimate rigidity to prevent underbonnet package violation.

Concept level system modelling is particularly useful at this stage in the design to pursue optimization and 'what-if' type studies to find ways of concurrently achieving each attribute. A typical example would be that of balancing road induced interior noise and vibration with vehicle handling precision, for a particular suspension design concept. In this situation the suspension bushes are required to provide both isolation and control. The NVH attribute requires suspension dynamic forces in the tactile and audible frequency ranges to be minimized in order to yield acceptable levels of noise and vibration. Meanwhile, the vehicle handling attributes generally call for maximum achievable bush stiffness in order to allow suspension systems to behave as jointed mechanisms.

The bushes alone do not control the handling or produce noise and vibration; they are but one component in the causal chain. They do however perform conflicting roles for the two attribute groups. It would be an oversimplification to state that NVH attribute requires flexibility and the dynamics attribute demands stiffness, when in fact each require particular values to ensure correct function. For example, the dynamicist may require a particular set of bushes to augment passive handling responses (e.g. lateral acceleration related understeer), whereas the NVH engineer will specify values that promote critical modal alignments (e.g. avoiding frequency coincidence between resonances of suspension and those of vehicle bodyshell). In this particular example it is usually necessary to adjust concurrently the performance of other independent components in order to make the common parts (the bushes) perform acceptably for both attributes.

The solution to too much road noise being transmitted via a particular bush may lie in reducing the bush stiffness. Alternatively, the force targets cascaded to the bushes may be relaxed if they can be balanced by more demanding body sensitivity targets (e.g. lower sound pressure at the occupant ear positions per unit bush force). In exceptional cases, increasing stiffness may change a resonant frequency such that interior noise is reduced. In most cases the dynamic force spectra are such that 'blanket' stiffness targets are inappropriate. There will be transmission paths where suspension geometry and road forcing effects combine to dictate that high bush stiffness is not a problem, and vice versa.

The model architecture at the early concept stage, during which the above target and attribute balancing exercise will be undertaken, is usually simple but contains enough detail to get the job done. Many parts will be represented by their cascaded target parameters, whilst components with some degree of definition or known physical form may be modelled in 3-D. Typically, the vehicle NVH model may be defined as follows in Table 1:

Table 1: NVH Suspension Subsystem Model – Road noise and vibration

Subsystem Model	Idealization
Hubs, wheels and tyres	Boundary: • Acceleration measured on target vehicles, or • Lumped unsprung mass properties on parameterized tyre properties
Suspension links	Physical links representing concept geometry, but with target stiffness and mass
Damping elements	Linearised scalar dampers
Bushes	Scalar springs / transfer functions constrained by target max / min static and dynamic values
Driveline and mountings	Rigid bodies attached to the vehicle model via scalar springs and dampers
Body	Body Model with adjustable parameters (see Figure 3) plus Target noise and vibration transfer functions
Result: Total interior noise and vibration from road surface induced hub acceleration/or road profile derived displacement spectrum/transient, with ranked contributions from all suspension/body interfaces	

In this example, measured hub acceleration (or road profile) data is used to excite a concept level physical suspension system model, which in turn drives an idealization of the vehicle body comprised of a matrix of noise and vibration transfer functions. This is a typical real world case, where models are comprised of benchmark test data, conceptual geometry, and cascaded target values.

9   SUBJECTIVE EVALUATION AT THE CONCEPT STAGE

Benchmark vehicle tests are often used in the automotive industry in an effort to 'short-cut' the above process of generating valid targets. This would appear to represent an attractively cost effective alternative, only requiring that the new vehicle exhibit lower noise and vibration levels than any competitor or predecessor. Two shortfalls of this approach are that firstly this in no way includes the brand values in the targets, and secondly this can be extremely wasteful of effort, in that the benchmark results may well be significantly below what a jury would judge (in a suitable context) to be the level of perception. An example of this is illustrated by figure 4 (1).

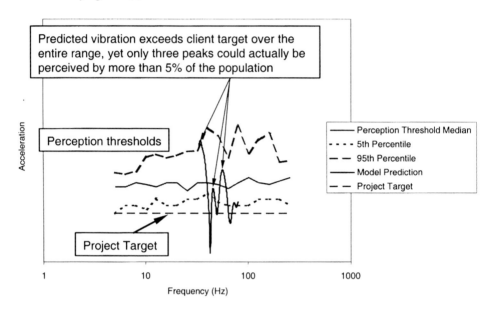

Figure 4: Perception thresholds vs. Benchmark Targets

Whilst automotive vehicle dynamic and NVH simulators are not yet commonplace, they are being developed and will become an integrated part of the concept assessment toolkit. Perhaps the most important aspect of any subjective assessment is that it should be made in the correct context, because the effect of acceptable 'background' or masking noise and vibration can disguise otherwise noticeable phenomena (1). Ideally all subjective evaluations will include noise and vibration, in an environment where the subject is performing tasks similar to the real world assessment conditions, i.e. driving a car. However, in the absence of a facility that enables this, at the least each stimulus should be assessed in its relative context. For example:

- Engine noise predictions need to be listened to at least against a background of road and wind noise.
- Low frequency (e.g. sub 250 Hz) road noise predictions need to be listened to in the context of the whole of the remaining target road noise spectrum.

C606/024/2002 © IMechE 2002

- Transient noise predictions (e.g. 'bump-thump' from small irregularities such as 'cats-eyes') need to be experienced with vibration. In this case, the interaction between sound and vibration is so strong that assessment of the sound or vibration in isolation can give misleading results.

The transient noise prediction shown in Figure 5 indicates the degree of difficulty involved in attempting to judge the difference between two sounds by inspection. The peak levels and decay envelopes do not appear to differ much between the two predictions, yet they sound quite different.

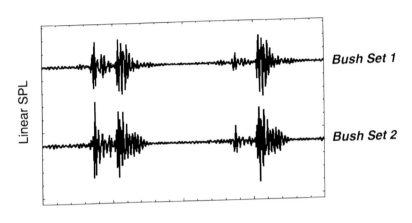

Figure 5: Transient Response Comparisons

In the case of transient response, the same was found to be true for the vibration signature. A metric has been developed for transient response through jury evaluation using a vibration test rig (1) (Figure 6). The resultant metrics generated by this integrated approach are often compound, in that there is usually a high degree of interaction between the responses to the various inputs to the assessor, including the sound of the transient. The effects of vibration phase are often important too. Antiphase (single frequency dominated) vibration between the human tactile interfaces (e.g. seat and foot lateral response) can lead to an impression of vehicle flexibility, whereas in-phase vibration (even at higher levels) can cause the opposite i.e. a feeling of well-crafted solidity. In this case a metric based upon the combined vibration magnitudes alone would be misleading.

C606/024/2002 © IMechE 2002

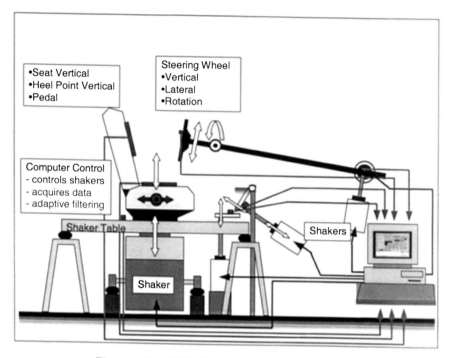

Figure 6: The MSXI Human Vibration Comfort Rig

Although the authors would advocate that the best results can be achieved when the interactions are assessed in as realistic an environment as possible, the rig referenced above has yielded useful target setting data for several automotive projects. MSXI are currently developing a Vehicle NVH Simulator to improve the environment in which subjective assessments of design alternatives, amongst other NVH issues, are made.

10 CONCLUSIONS

The process described in this paper has been used by the authors to enable the Brand Values to drive the design of a new vehicle, through cascaded, validated targets. Sub-system and even component targets can be balanced using concept analysis very early in the program, and these targets can be validated by re-synthesis into physical stimuli, which can be assessed subjectively in the correct context. The building blocks to perform the process are, for the most part, readily available.

11 REFERENCES

[1]   Dr. Joseph Venor. "Comfort Assessment and Vibration Quality as Part of the Vehicle Design and Development Process". IMechE Total Vehicle Technology International Conference, Sept. 2001.

C606/024/2002 © IMechE 2002

C606/014/2002

Designing for Six Sigma reliability

D BRUNSON and **B HALLAM**
Jaguar Cars Limited, UK
S MISTRY
Tesma Engine Technology
I F CAMPEAN
School of Engineering, University of Bradford, UK

ABSTRACT

This paper discusses the reality and definitions of delivering six sigma reliability in complex automotive systems and products using a case study to illustrate the concepts.

It is argued that the numeric interpretation of six sigma is less important than process rigour, which can be applied to engineering design by using combinations of statistical and computer aided analytical methods. The continuous improvements in reliability available from such methods can deliver complex systems with high reliability. The economic and resource factors of applying varying levels of engineering integrity across large projects are also considered.

1. INTRODUCTION

The usual definition of a six sigma capable process (long term) is one that achieves 3.4 defects per million opportunities (1). In terms of delivering reliability, this would translate to 3.4 failures or faults/million opportunities for the total mission of a system, which is 10 years or 150,000 miles for a typical modern car. Currently, many automotive companies are achieving 20 failures/car in a ten-year life. Assuming that there are an average of 1000 components/car, and each component has the opportunity to fail, this gives 20000 failures/million/component i.e. 3-4 sigma. Considerable improvements to the design processes are required to achieve the improvements in reliability needed to approach failure rates associated with six sigma.

2. THE PROBLEM WITH ACHIEVING SIX SIGMA RELIABILITY

2.1 Probabilistic perspective

3.4/1000000 * 1000 = 3.4/1000 = 99.9966 % reliability for a 1000 component product i.e. the typical probability of failure for a component needs to be 0.0000034 over 10 years or 150,000 miles, which is in the order of 10^{10} hours per failure. Such a reliability level is normally only encountered in nuclear and aerospace applications, and it is obviously an unreasonably small failure rate to expect to verify prior to volume production beginning for an automotive application.

A more realistic target in "sigma" terms would be to aim for a 1 sigma improvement in reliability level. This would mean 2000 failures / million opportunities, i.e. 2 failures / car / useful life. This implies that at component level the failure rate should not exceed 0.002 (i.e. in the order of 10^6 hours per failure) and 99.8% reliability for the useful life.

2.2 Consumer Perspective

The probability of getting a "perfect" product over its whole life can be defined in terms of the customer requirement of "no faults when new and no deterioration in prime function with age". Using a strategy of managing faults such that if they occur they have the minimum effect on consumers will enable 99.8% reliability levels to achieve high customer satisfaction.

Even so, reliability demonstration to such high levels needs alternative techniques to traditional automotive reliability demonstration methods (2).

Requirements to prove such high levels of reliability before high volume production begins requires both *measurement* and *management* of reliability, not just measurement of reliability using traditional technical tools or just relying on "good management" of product development to achieve high levels of reliability.

3. DESIGN PROCESS SOLUTIONS

In order to achieve a substantial increase in product reliability the design process itself must be scrutinised. Existing quality systems are based around achieving consistent output of product from manufacturing, but only a few are currently attempting the same for the total design process (the product development factory as it is sometimes called). The success of Six Sigma as a problem solving and process improvement initiative is well documented (1)(3) but the extension of the philosophy and a practice into product design is less evident. There are enough similarities between the aims and methods of Design for Six Sigma (DfSS) and Reliability and Robustness Engineering methods to propose an approach that combines them. As shown in the diagram in Figure 1 the Design for Six Sigma approach breaks the design process into four distinct stages:

C606/014/2002 © With Authors 2002

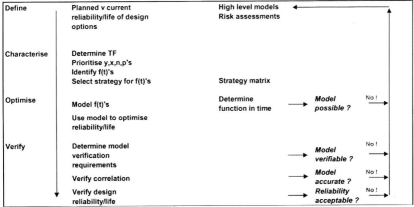

Figure 1. DfSS reliability process flow

- Definition
- Characterisation
- Optimisation
- Verification

Each of theses stages has elements where product reliability needs to be considered.

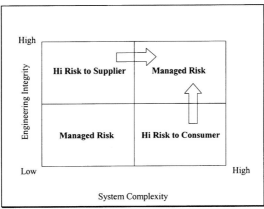

Figure 2. Design strategy diagram

3.1 Definition

The purpose of the definition stage is to ensure that design resources are allocated in the correct proportion to the correct areas of design, so that their utilisation is optimised. The strategy diagram shown in Figure 2 indicates a straightforward means of determining this. The engineering effort required to achieve high integrity design needs to be applied to the areas of design where the greatest reliability risks to the customer is evident. The DfSS tools will help with achieving the level of detail required to deliver high reliability for complex products. However, if applied where unnecessary, the additional resource required to apply the in depth analysis tool could be wasted. Efficient design planning and analysis tools need to be utilised to position each technology or system within the strategy map and apply the appropriate level of design effort. Employing tools such as Quality Function Deployment, Failure Mode and Effect Analysis, Fault Tree Analysis, Reliability and Robustness Specification processes and combined Design Functional Specifications, and Function Structure Trees, could generate means of quantification to achieve this level of definition. Such tools generate most benefit when used at all levels of the project, e.g. total vehicle level cascading down to individual components and processes in the case of a car (4).

3.2 Characterisation

The aim of the Characterisation phase is to enable the design process to identify and manage all significant design parameters (in terms of both nominal and variation) and their combinations. The means of achieving this is to develop a Transfer Function (TF) using a simplified regression approach of the form:

$$Y = f((x1, \sigma x1)t, (x2, \sigma x2)t, (x1.x2)t\ldots\ldots).$$

Once developed, the Transfer Function can be also used as a metric to quantify design complexity, the number of factors and terms involved indicating the complexity of the total design. If fully quantifiable, then the TF can make design optimisation straightforward. However, in practice it is most likely that the TF will be only partly quantifiable, therefore for other factors, risk management methods (Design Failure Mode and Effects Analysis (DFMEA) must be applied. DFMEAs help to ensure that similar designs are capable from current data and documented evidence and data is available to prove that new design can only move in a positive direction.

The TF elements that have a time related characteristic must be treated using reliability methods. This will often be a small number of parameters in the total TF, which reduces the complexity of the model. Functions that explain time related deterioration are often complex so such in depth analysis must be applied strategically to conserve resources. Use of Noise Factor Management methods can determine where more complex time related functional deterioration needs to be modelled. This technique is similar to that used in the Definition stage to ensure the correct tools are applied to the appropriate commodity areas to be efficient in the use of engineering resource. Current experience of the DfSS process is that much of the TF is not fully understood when decisions that affect the reliability of the product are made. The reliability related sub-functions inevitably need to be investigated early in the programme because the proving of the correlation takes more time than with functional relationships.

A major advantage of developing complex models early in the design process is that they can often then be applied generically to other similar areas of design.

3.3 Optimisation

The decomposition into design parameters and development of multiple regression models, which provides the essence of the above characterisation process, allows structured investigations and experimentation using statistical methods to generate models, which represent functional and reliability performance. The models developed from such structured investigations enable processes for finding optimal solutions for both means and variability of reliability performance simultaneously. There are a number of computer based optimisation methods that enable engineers and scientists to achieve complex mathematical optimisation with relative ease (for example optimisers using Successive Linear Approximation Methods, Minitab Solver and Excel Solver). These techniques can be combined into processes that will allow multi-optimisation of all consumer critical attributes. Computer modelling such as Computational Fluids Dynamics (CFD) and Finite Element Analysis (FEA) combined with statistical methods such as design of experiments and multiple regression, allow effective optimisation to be achieved where traditional computer aided models that ignore noise factors, interactions and variability provide compromised overall product performance. The optimisation methods are used in effect, to reconstruct the all elements that make up the parameter space, in order to investigate the areas that provide the most robust and reliable solutions.

3.4 Verification

Design verification methods traditionally consist of qualification of reliability based on small sample size demonstrations using prototype components, manufacturing processes and design features (2). The statistical value of such demonstrations is dubious, but it is often the only data available to engineers and managers prior to fully manufactured products being available.

Following the DCOV process allows verification to be based on quantification of the reliability performance predicted by computer based models. Verification techniques based on these principles can provide much greater levels of information to predict field reliability performance of the final product. Moving from qualification towards quantification by using a combination of statistical and analytical methods fits the Six Sigma philosophy of using data to drive decisions on characteristics that are of most importance to the customer or consumer. This can be achieved by careful planning of design verification activities to gather the significant data on the most representative characteristics and correlating the results from models and the physical measurements and results gathered during design verification testing.

In reality many of the computer-based models require tuning once verification measurements and results are gathered. The iterative nature of the total process means that much of the work needs to start early in the design process. Since the overall process concentrates on applying the maximum engineering rigour to areas of design that are most critical to the consumer, this early verification activity need not be extensive. This adds considerably to the effectiveness and efficiency of the design verification process.

4. COMBINING STATISTICAL AND CAE TOOLS

A major contribution to the effectiveness of the DCOV methodology is derived from the combined use of statistical methods and Computer Aided Engineering (CAE) tools. The majority of CAE work uses models deterministically. For high volume products statistical thinking reveals the need to determine the customer percentile represented by a models output. If mean values are used the percentile will be 50%. This is obviously an inadequate representation of the customer base as only half of the product population is being represented by the model. This may be an extreme example but the authors have seen much evidence of it happening. Many CAE models are aimed at simulating "worst case" scenarios that are used to generate "safety factors". Such analyses can lead to over design if the contribution of the various factors is not understood. Beefing up the design hardly justifies the use of sophisticated and resource intensive CAE tools. Using the tools to optimise designs from development of an increased understanding of the stochastic nature of the design space will be more beneficial to both the manufacturer and the consumer.

These issues require a change in the stage at which resources are typically utilised during the design process in the automotive industry. Management, and especially project management, tools are therefore required.

Historically, individual engineers would be able to design a system from concept to the marketplace, but the complexity of today's automotive systems make team approaches to product design essential. The communications and data management requirements of working in diverse teams leads to the use of a set of tools usually labelled under "quality engineering". These range from Quality Function Deployment to Design Verification Plans (5) and generally add value by providing more rigour and control to the engineering processes.

There is no mention of "quality management" so far in this discussion because it is suggested that engineering management methods need to change to de-bug systems and components as they are being designed to remove the source of errors. The additional effort required to add the rigour is justified by savings in effort later in the design process where traditional "find and fix" methods are applied. Comparing this with a manufacturing scenario Statistical Process Control (SPC) takes data from a process, translates it to information via analysis using statistics and communicates the information to an earlier stage in the process where the effect of change is more efficient. Engineering management need to achieve the same for the design process.

A prime example of where such improvements can be made is the use of statistically designed experiments to allow much greater understanding of the potential effects around the design space (6). Traditional techniques only allow investigation of a small area of the design space, so optimisation is seldom achieved. In fact, designs are often compromised not optimised. The design and analysis process needs to put all known factors into the design space early in the process and optimisation methods applied. Some surprises are still possible but the probability of them being the most influential factors is reduced if all factors are being considered.

Such systems engineering methods and tools are an enabler to achieving efficient models and can be used along with analytical and Computer Aided Engineering methods to achieve very high levels of reliability (7).

Stochastic modelling is possible and getting practical. However, it does require statistical design methods and data on all influential design factors to be accurate. The amount of effort that is required to collect factor data, surrogate or real, is currently underestimated in auto industry. Such data is often collected too late to influence the design, which cause 70 to 80 % of failures/customer concerns.

Such models require calibration and verification and this can become the principal direction for prototype testing in place of traditional "sign off testing".

C606/014/2002 © With Authors 2002

5. CASE STUDY

The following case study is presented to illustrate the application of the solutions above in a real design scenario. It gives an example showing how the integration of analytical and statistical tools into the design process at various stages has been utilised to enhance product reliability. Discussion based on the case study evidence will also consider the organisational, management and financial implications of delivering increased reliability using such techniques.

This case study considers a gasket used to provide a sealing function between a pipe and an internal combustion engine cylinder block. The transfer function identifies only two factors that will vary in time and hence affect the reliability of the system. The TF in this case is of the form:

$SL(t) = ICL - ES(t) - CLL(t) - (ES.CLL)$

The function at the end of life has been estimated by establishing a model with conditions with each parameter represented in an aged and new condition in a Finite Element Stress analysis (Figure 4) built up from time dependant linear relationships of clamp load loss (CLL) and elastomer stiffness (ES). The only time independent factor in this case being the initial clamp load (ICL). The overall model needs to consider the interaction between two elements, but because they are both represented in the aged condition the end of life function can be established. The end of life function is expressed as a factor of the seal load and the total stress in the seal at which leakage will occur with all variation included. As shown in Figure 3 the prediction with no variability indicates no leakage with a seal load as low as 5N/mm, but when variation is included (Figure 5) the proportion of failures starts to increase above 13N/mm. The finite element model in Figure 6 shows the predicted area of leakage on the sealing face.

To correlate the modelling results with physical measurements a rig test was run using prototypes with similar attributes set at limit conditions equivalent to those at which the model predicted seal leakage would occur.

The results of the test revealed seal failure after 150 hours of accelerated testing (correlated to be equivalent to full vehicle life for sealing damage accumulation). Figure 7 shows the damage incurred at the seal face accumulated over the test. The area of low seal load is clearly shown to the right of the sealing face. The angular difference between the predicted and the actual leak area is explained by the difference in geometry of the surrogate pipe and the design intent indicated by the model in Figure 3.

The seal failing on test as predicted by the model verifies the correlation between model and test results. Further model runs at varying parameter levels can now be conducted to determine the sensitivity of the design to the time related parameters on the accelerated test to develop generic design rules that include the effects of variation.

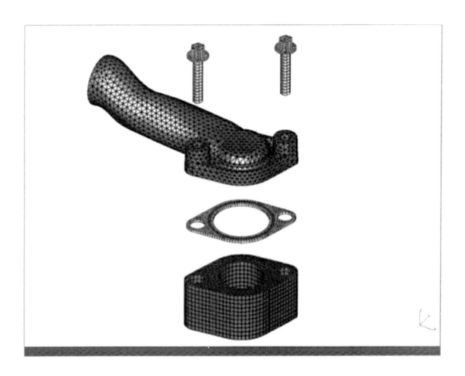

Figure 3. Finite element model of sealing system

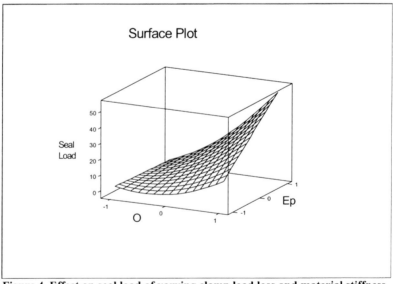

Figure 4. Effect on seal load of varying clamp load loss and material stiffness

C606/014/2002 © With Authors 2002

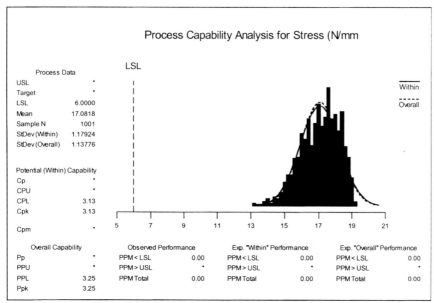

Figure 5. Variation in clamp load due to TF factors

The case study provides transfer functions, that can be applied to future design alternatives to establish the degree of reliability achievable within similar ranges of design and process parameters.

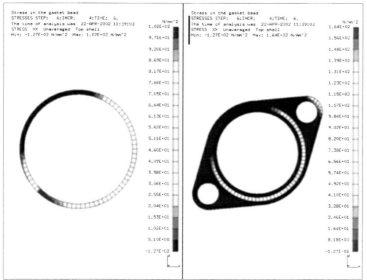

Figure 6. Predicted area of leakage under limit conditions.

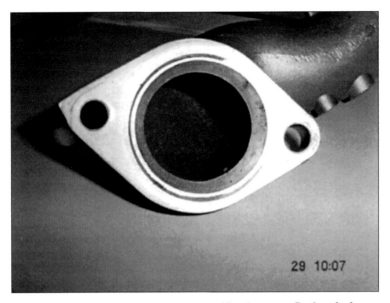

Figure 7. Result of limit condition verification test. Coolant leak.

6. CONCLUSION

This paper has discussed that high level product reliability can be achieved using a combination of statistical and computer aided engineering methods. A case study from a complex automotive system has been used to illustrate how the design process needs to be structured to maximise the reliability achievable from a given set of design parameters. Consideration has also been given to the commercial aspects of providing efficient resource use through the design process by analysing product quality at a strategic level.

REFERENCES

(1)   Breyfogle III, F.W., 1999, *Implementing Six Sigma, Smarter Solutions Using Statistical Methods.* Wiley-Interscience.

(2)   Ireson, G.W. and Coombs C.F.Jnr., 1988, *Handbook of Reliability Engineering and Management.* McGraw-Hill.

(3)   Eckes, G., 2001, *Making Six Sigma Last, Managing the Balance Between Culture and Technical Change.* John Wiley and Sons Inc.

(4)   Edwards, S.P., Grove, D.M., Wynn, H.P., 2000, *Statistice for Engine Optimisation.* Professional Engineering Publications.

(5)   Brunson, D., 1998, *The Development of a Model to Improve the Reliability Proving Process.* MSc Thesis University of Bradford.

C606/014/2002 © With Authors 2002

(6)   Moen, R.D., Nolan, T.W., Provost, L.P., 1991, *Improving Quality Through Planned Experimentation*. McGraw-Hill, Inc.

(7)   Campean, F., Day, A.J., Wright, C.S., 2001, *Camshaft Timing Belt Reliability Modelling*. Proc. Annual Reliability and Maintainability Symposium.

C606/016/2002

Dimensional variation analysis for automotive hybrid aluminium body structures

R KOGANTI and **M ZALUZEC**
Ford Research Laboratory, Ford Motor Company, Dearborn, Michigan, USA

ABSTRACT

Ford Motor Company is investigating various lightweight material systems targeting improvements in both vehicle performance and safety of occupants during vehicular collision. Aluminium is often the first choice of lightweight materials for improved fuel economy due to the materials current usage for non-structural applications. More recently, aluminium is being investigated for structural development as well due to its energy absorption characteristics. Aluminium structures have many advantages in automotive applications due to its performance in crash applications. Although unitized steel body architectures represent the most common body structure material system, aluminium architectures can come in a variety of designs including unitized body, space frame and hybrid body architectures. Stamped and extruded (straight and hydroformed) components are finding more applications in unitized body architectures as automotive engineers find cost savings associated with applying these types of aluminium products. Hybrid aluminium architectures consist of a combination of aluminium products including stampings, extrusions (straight and hydroformed) and castings. Joining and assembling these components is often a challenging task, since most of the automotive and joining infrastructure is based on spot welded steel assemblies. In the case of aluminium, a multitude of joining methods can be applied to join extrusions, stampings, and castings. Commonly used techniques include rivets, MIG welding, laser welding, and adhesive bonding.

The development of manufacturing processes for joining and assembling of lightweight aluminium vehicles requires detailed process capability studies as well as dimensional variation analysis (DVA) studies to ensure process controls are in place. These manufacturing processes not only have to provide cycle time viability but also need to maintain or surpass product safety and quality. Initially to understand the variation of body structures a DVA was conducted on surrogate hybrid aluminium structures considering both component and fixture tolerances. Using Monte-Carlo simulations, final body assembly structure variability and process capability were predicted. DVA gives a roadmap for the subsequent product design and analysis in aluminium structures development work to ensure the highest quality standards in next generation vehicle body construction.

1 INTRODUCTION

Customer satisfaction is the primary emphasis Ford Motor Company and others is placing of its vehicle products. To enable customer satisfaction, Ford Motor Company is implementing the six sigma philosophy across all corporate levels. The goal is significantly to improve customer satisfaction of the Company's products and services by significantly reducing variability and defects in manufacturing and engineering operations. Successful implementation of Six Sigma philosophy will yield only 3.4 defects per million parts [1].

"Dimensional management" is a key aspect in body structures in the design phase. Automotive body structures involve multiple parts, fixtures and various joining manufacturing processes. Variation of parts or processes can cause a significant impact on body structures dimensional management and ultimately an adverse impact on customer satisfaction. Variations of a door assembly can lead to poor sealing, wind noise, water leaks and poor craftsmanship. During the design phase the geometrical tolerance of each component and its influence on assembly is addressed.

This paper discusses the current manufacturing research on advanced metal inert gas (MIG) welding for joining aluminium components in body construction, and the impact of dimensional Variation Analysis (DVA). DVA is a part of Design For Six Sigma (DFSS) methodology to identify the bottle neck points in the design phase so that manufacturing process meets the six sigma standards when the product is launched.

In order to achieve the six sigma process capability, the product needs to be designed to achieve the target process capability. Defect elimination not only improves customer satisfaction but also results in reduced warranty costs. DFSS is an efficient tool to be considered in the design phase for any new manufacturing process to achieve the target process capability. In body architectures the variation in dimensions in body structures could lead to product performance variations. For example if two attributes in body architecture have variations, the final product performance will vary due to the variations of the two factors as shown in figure 1.

This paper will focus on the application of DVA in dimensional management of hybrid aluminium body structures for automotive applications. The objective of the project was to achieve dimensional process capability of 1.67 for the hybrid front end module using MIG welding joining process.

C606/016/2002 © IMechE 2002

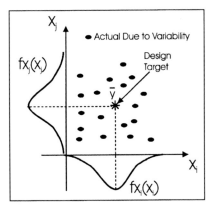

Figure 1: Product performance variations due to process variations

2 HYBRID FRONT END STRUCTURE ASSEMBLY DESIGN

Ford Motor Company is investigating various light weight materials for high mileage performance and also safety of occupants during collision. Aluminium is one of the chosen materials for the structural development work for high mileage as well as crashworthiness. Aluminium structures have many advantages in automotive applications due to its light weight and energy management performance in crash applications. Various automotive manufacturers are currently using aluminium in structural applications. The Plymouth Prowler, Audi A8, Audi A2, Ferrari's 360 Modena, Panoz's Esperante, Ford's - Th!nk Neighbor, and Honda's NSX and Insight, are just a few of aluminium intensive vehicles currently in production. Although unitized steel body architectures represent the most common body structure, aluminium architectures can come in a variety of designs including unitized body, space frame and hybrid body architectures. Hybrid laser/MIG welding, magnetic pulse welding, and friction stir welding are future enabling technologies in aluminium joining processes for hybrid architectures. The Ford's Scientific Research Laboratory (SRL) is currently investigating the joining and assembly of hybrid aluminium body architectures using MIG welding as the primary joining process. One of the main attributes in any body architecture is dimensional management of Body-In-White (BIW) assemblies. Maintaining tolerances from individual components to final assemblies is complicated. Besides individual tolerances of components and assembly fixtures, other sources of variation that influence dimensional process capability is process variation. However, current DVA models only evaluate in design phase is stack-up tolerances based on individual component and fixture tolerances. DFSS is applied in the design phase of new products to identify the bottle neck points and to improve the ultimate dimensional process capability. The details of dimensional process capability are explained in the subsequent section. The hybrid aluminium front end assembly use in this study is shown in figure 2. The manufacturing methods employed to components with the hybrid front end assembly are shown in table 1.

Table 1: Front end module components manufacturing process details

Mfg. Process	Components
Sand Casting (Aluminium)	Shock absorber towers
Extrusions (Aluminium)	Radiator support Bumper beam A-pillar inner Crash Box Longitudinal Rail Top cross-member H-bracket Shock tower support Vertical rail
Stampings (Aluminium)	Torque box upper & Lower Reinforcements Brackets for vertical rail support

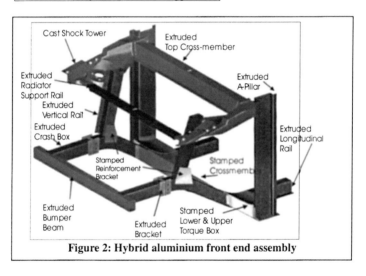

Figure 2: Hybrid aluminium front end assembly

3. DIMENSIONAL PROCESS CAPABILITY

For a typical high volume manufacturing process, the main causes of variation can be due to people, machines, materials, methods, environment and measurement [2]. If any of these elements change, the process system will be changed and, hence, there will be corresponding change in product quality. The variations in the process can result in variations in the end product's performance. The process capability indices, process potential (P_P) and process performance (P_{PK}), are the units of measurement for the process robustness. Process potential

C606/016/2002 © IMechE 2002

can be defined as the range within which the majority of the parts or values in a distribution will lie. The process performance will be used in the conceptual phase.

The process performance for the assembly dimensions are calculated based on the formula:

$$P_P = \frac{\text{Specification Tolerance}}{\text{Process Capability}} = \frac{USL - LSL}{6\sigma}$$

Where USL and LSL are the upper and lower specification (tolerance) limits; these specifications vary from application to application.

In the pre-production phase, in addition to process potential values, the performance of a process (P_{PK}) relative to design specification should be determined. Process performance considers the effect of the actual mid point of the process relative to the midpoint of the specification. The P_{PK} can be calculated as follows:

$$P_{PKUpper} = \frac{USL - \mu}{3\sigma} \quad \text{or} \quad P_{PKLower} = \frac{\mu - LSL}{3\sigma}$$

Where μ is the process average and σ is the standard deviation of the measurement. The lowest value out of $P_{PKUpper}$ or $P_{PKLower}$ is the measure of the process performance. P_P and P_{PK} values, in manufacturing language, relative to parts out of specification, are as shown in table 2.

Table 2: P_P and P_{PK} values and number of parts out of specification

P_p & P_{pk}	% non-conforming	% yield	# of parts out of specification
1.0	0.2700	99.73	2.7 parts per thousand
1.33	6.3342×10^{-2}	99.9937	63.3 parts per million
1.67	5.7330×10^{-5}	> 99.9999	573.3 parts per billion
2.0	1.9732×10^{-7}	> 99.9999	2.0 parts per billion
3.0	2.2572×10^{-17}	>99.9999	0.0 parts per billion

Automotive Industrial Action Group (AIAG) is a North American automotive consortium formed by Daimler-Chrysler, Ford Motor Company and General Motors to provide an open forum where members cooperate in developing and promoting solutions that enhance the prosperity of the automotive industry. Quality operating procedures and standards for automotive is one among other standards (purchasing, finance, materials management, project management, etc..) developed by AIAG. According to the AIAG quality standards, the P_P and P_{PK} should meet the requirement of ≥ 1.67 in the part-production approval process (PPAP) phase [3]. When the part goes into the production phase, the P_P and P_{PK} should meet the requirement of ≥ 1.33. Since the MIG welding development work is in prototyping phase, the goal was to achieve a P_P of ≥ 1.67.

The relationship between P_P and P_{PK}, and the process shift in a dimensional capability study, is shown in figure 3. It can be seen that even though the P_P achieves 1.67, in all cases, P_{PK} can be less than 1.67 when the process shifts from the target. This would then necessitate process improvement, through process control or modifications to the tool or fixture, in order that in the pre-production and production phases, a P_{PK} of 1.67 can be achieved. It should be noted that P_P represents the process consistency, whereas, P_{PK} represents not only consistency but also the design target.

The mean and standard deviations of the points selected for the analysis are calculated using Monte Carlo simulation. The simulations details and methodology is explained in the subsequent section.

Figure 3: Relationship between P_P and P_{PK}

4 DESIGN FOR SIX SIGMA (DFSS) IN DIMENSIONAL MANAGEMENT

Dimensional management is an important attribute in automotive structures. Inadequate dimensional management in automotive structures can lead to customer dissatisfaction specific to poor craftsmanship, poor sealing, squeaks and rattles, and NVH (noise, vibration and harshness).

The objective of DFSS is to identify the bottleneck stack up tolerances to maintain acceptable tolerance for the final assembly. In each phase, starting from individual components through final assembly, a certain tolerance is allocated to calculate the overall process capability.

In the design phase, typically dimensional variation analysis (DVA) will be conducted from individual component level to final assemblies. The following steps are conducted in DFSS phase to evaluate the dimensional P_P and P_{PK} of body structures

Step1: Define the part and fixture tolerances
Step 2: Define the assembly process sequence
Step 3: Conduct DVA study to calculate Potential P_P and P_{PK} and also identify the potential stack-up tolerances

The subsequent sections will discuss the specific on tolerances and DVA study.

4.1 Tolerances for hybrid aluminium components and assembly fixtures
The tolerances are important information needed for the dimensional variation simulation. Component tolerances are established based on the historical process data. Since the

aluminium front end consists of extrusions, stampings and castings, each aluminium product has different tolerances based on component design and manufacturing methods. The summary of tolerances for extrusions, stampings and castings are given in table 2. Besides the components tolerances, the other critical tolerances needed for simulation is the assembly fixtures tolerance. The individual and sub-assembly parts are constrained in the assembly fixtures using net pads (also referred as datums) and pins. These parts are clamped into place to ensure the parts are in contact with the datums during the joining process. The tolerances adopted for individual components and fixture details are given table 3. These component and fixture tolerances are fed into DVA model for estimating the final assembly tolerances and dimensional process capability.

Table 3: Individual component and fixture tolerances

Mfg. Process for Components	Tolerance Specification
Casting	Mating surface: + /- 0.75 mm Machined surface +/-0.50 mm
Extrusion	Mating surface: +/- 0.25 mm End cuts: +/- 0.5mm Hole location: +/- 0.5mm Stretch bent rail surface: +/- 0.75 mm
Stampings	Mating surface: +/- 0.7 mm

Assembly Fixture Tolerances:	
Net Pads	+/- 0.125 mm
Pin Position	+/- 0.125 mm
Pin Size	+ 0.00 and –0.03 mm

4.2 DVA using Monte Carlo simulation (MCS)

Monte Carlo analysis is a powerful engineering tool by which one can perform a statistical analysis of the confidence level (or uncertainty) in engineering problems. It is particularly useful for complex problems where numerous random variables are related through non-linear equations.

The basic principle in Monte Carlo simulation is the generation of a set of random numbers for the response variable. The magnitude of dimensional variation can be found by performing a large number of replications in random manner by standard software available from the computer subroutines library. The advantage with the Monte Carlo is that even if the individual variables follow non-normal distributions, the final output will result in normal distribution.

Furthermore, the dimensional variation ranges for the critical points were obtained by using VisVSA Software [4]. The variations obtained by Monte Carlo method are calculated for 2000 replications and can be seen in table 4. Basically, the higher replications will yield better accuracy of the response variable. In design phase all the part and assembly fixture tolerances are inputted into the DVA model, and final assembly inspection points will be assessed. A total of 24 dimensional characteristics were identified as critical for subsequent assembly operations.

An assembly tolerance (upper and lower specification limits) of +/- 1.5 mm is applied for all critical points in the DVA analysis.

5 RESULTS

A total of 24 significant "critical characteristics" (dimensions deemed very sensitive or critical to the achievement of fit, function and finish) were analysed for dimensional process capability at the design phase. These dimensional points are critical points in the subsequent assembly process. Monte Carlo simulation was used to analyse the dimensional process capability. Using the part and fixture tolerances (table 3), the final dimensional process capability in the design phase is calculated. The probability density function (PDF) for the dimensional variation of a critical point 207_Hx and sensitivity analysis are shown in figure 4 and table 5 respectively. Based on the simulation results, the total hole location of the shock tower can vary +/- 0.30 mm (3 standard deviations) and short term P_P and P_{PK} are 4.46 and 4.45 respectively. These results only include part and fixture variations and this doesn't include process variations. Figures 5, 6, and 7 show the locations and variability (+/- 3 standard deviations) of the points on the front end structure. The statistics of all the dimensional points are shown in table 6.

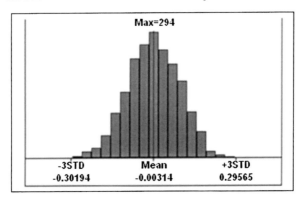

Figure 4: PDF for DVA for measurement point 207_H-x (left hand shock tower hole location)

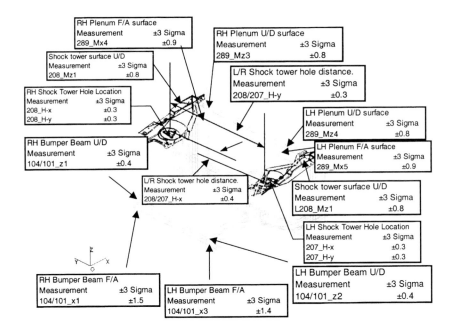

Figure 5: Measurement points on bumper, shock tower and plenum member

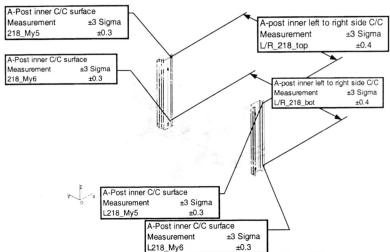

Figure 6: Measurement points on left and right side A-Pillar

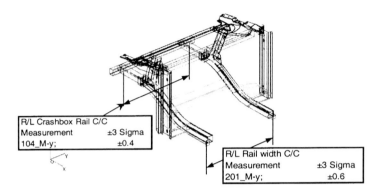

R/L Crashbox Rail C/C	
Measurement	±3 Sigma
104_M-y;	±0.4

R/L Rail width C/C	
Measurement	±3 Sigma
201_M-y;	±0.6

Figure 7: Variability assessment at crash box and longitudinal rails

Note: All dimensions shown in figures 4-7 are in millimeters

Table 4: Statistics Summary of MCS data (for measurement point # 207_H-x)

Trials	2000	Mean	0.00 mm
Nominal	0.00 mm	Standard Deviation	0.099 mm
P_P	5.02	P_{PK}	5.01
Range of Minimum	-0.31 mm	Range of Maximum	0.379 mm
Range Width	0.69 mm		

Table 5: Sensitivity analysis of % contributions of sub-assembly fixture tolerances on left hand shock absorber tower hole location (for measurement point # 207_H-x)

Index	Tolerance description for Front End Sub-Assembly Fixture	Tolerance +/- (mm)	Percent (%)
1	Front end sub fixture pin location tolerance	0.125	28.00
2	Shock absorber hole pin clearance	0.140	28.00
3	Shock absorber pin location tolerance	0.125	22.30
4	Shock absorber tower net pad tolerance	0.125	11.00
5	Shock absorber tower net pad tolerance	0.125	10.10

C606/016/2002 © IMechE 2002

Table 6: Dimensional Variation Analysis of Front End Module for all the critical points

Ref #	Average (mm)	Nominal (mm)	Std.Dev (mm)	6*Std.Dev (mm)	P_P	P_{PK}
208_H-x	0.00	0.00	0.11	0.65	4.59	4.57
208_H-y	0.00	0.00	0.09	0.55	5.43	5.43
207_H-x	0.00	0.00	0.10	0.60	5.02	5.01
207_H-y	0.00	0.00	0.09	0.55	5.47	5.46
208_Mz1	-0.01	0.00	0.26	1.54	1.95	1.94
L208_Mz1	0.00	0.00	0.27	1.60	1.87	1.86
289_Mz3	0.00	0.00	0.27	1.61	1.86	1.86
289_Mz4	0.00	0.00	0.27	1.61	1.187	1.87
289_Mx4	0.00	0.00	0.29	1.73	1.73	1.73
289_Mx5	-0.01	0.00	0.29	1.73	1.73	1.72
104/101_z1	0.00	0.00	0.13	0.77	3.92	3.91
104/101_z2	0.01	0.00	0.15	0.88	3.40	3.39
104/101_x1	0.01	0.00	0.49	2.95	1.02	1.01
104/101_x3	-0.01	0.00	0.46	2.76	1.09	1.08
218_My5	0.00	0.00	0.11	0.68	4.44	4.44
218_My6	0.00	0.00	0.10	0.60	5.00	5.00
L218_My5	-0.01	0.00	0.11	0.68	4.41	4.40
L218_My6	0.00	0.00	0.10	0.62	4.83	4.82
208/207_H_y	1098.41	1098.40	0.11	0.64	4.70	4.68
208/207_H-x	0.00	0.00	0.13	0.77	3.88	3.88
L/R_218_top	1354.20	1354.20	0.13	0.80	3.73	3.72
L/R_218_bot	1354.20	1354.20	0.13	0.80	3.73	3.72
201_M-y	881.68	881.68	0.19	1.17	2.57	2.56
104_M-y	881.69	881.68	0.15	0.88	3.42	3.41

6 SUMMARY

- Six Sigma philosophy improves the quality of the process by reducing the variability, and ultimately defects reduction up to 3.4 ppm
- Any existing or new products or processes, Six Sigma philosophies can be applied to reduce the variability and to improve the quality
- Six Sigma philosophy was applied for the application of MIG welded aluminium front end assembly
- DVA which is a part of DFSS methodology applied for the anticipation of stack up tolerances of hybrid front end components.
- Application of Monte Carlo method in dimensional management for automotive structures is discussed.

- Monte-Carlo method used 2000 replications to estimate the mean, standard deviation, process capability (P_P and P_{PK}) of all critical dimensional points on the front end assembly

7 CONCLUSIONS

- DVA is a part of Design For Six Sigma Methodology and it identifies the bottle neck points in the design phase
- DVA is a critical tool in design phase to assess the dimensional variations of the assembly.
- Dimensional Process Capability of 1.67 was achieved for the hybrid aluminium front end assembly using assembly tolerance of +/- 1.5 mm.
- The Dimensional Process Potential of assembly meet Ford's requirement of equivalent or greater than 1.67 based on assembly tolerance of +/- 1.5 mm.

8 RECOMMENDATIONS

- The process capability obtained from Monte Carlo simulation needs to be verified with front end assemblies measurement data.
- The process capability values need to be updated using the measurement data from the front end assemblies.
- The variability of process should be considered in the Monte-Carlo simulation for accurate predictions

9 REFERNCES

1. Mikel J. Harry, Richard Schroeder, Six Sigma: The Breakthrough Management Strategy Revolutionizing the World's Top Corporations, Published by Double Day, February 2000.
2. Gordon H. Robertson, Quality Statistical Thinking: Improving process control and capability, ASI Press, 1990.
3. Production Part Approval Process (PPAP), Second Edition, Guidelines developed for Chrysler Corporation, Ford Motor Company, and General Motors Corporation by American Society for Quality Control (ASQC) and Automotive Industrial Action Group (AIAG), February, 1995.
4. VisVSA Software Manual, EAI, 2000.

C606/026/2002

Reliability-based multidisciplinary design optimization of vehicle structures

R J YANG, L GU, and C H THO
Ford Motor Company, Dearborn, Michigan, USA
K K CHOI and B D YOUN
Center for Computer-aided Design and Department of Mechanical Engineering, The University of Iowa, USA

1. ABSTRACT

Crash Simulation evaluates vehicle crashworthiness for occupant safety. It is a highly undeterministic process due to uncertainty in every aspect, e.g., material, gages, etc. As a result, conventional deterministic optimization may be insufficient for crash problems. In addition, a system approach, which considers multiple crash modes, must be applied for safety, as a design considering only a single crash event may lead to failures in the other modes. Reliability-based multidisciplinary design optimization is one of the best ways to address uncertainty problem efficiently. It provides an analytic and systematic tool for considering uncertainty in product development process. However, it is often infeasible due to the lack of computational resources, simulation capabilities, and efficient optimization methodologies, as reliability analysis itself is a design optimization problem that requires many function evaluations. This paper investigates two reliability-based optimization methodologies: reliability index approach and performance measure approach. They are applied to a full vehicle design of multiple impact modes including full frontal impact, roof crush, side impact, 50% frontal offset impact, and other potential safety performance measurement, with uncertainties taken into consideration. The results of these two approaches are compared and discussed in detail. It is concluded that the performance measure approach is more efficient and robust than the conventional reliability index approach.

2. INTRODUCTION

Continuous demands for efficient designs on vehicle safety, NVH, durability, and other attribute performances have placed increasing emphasis on the analysis of the vehicle structural behavior as well as occupant behavior. Numerical methods have been widely used for this purpose. The most popular and flexible computational methods for vehicle design are the finite element methods. Over the past ten years, tremendous increases in computer speed and rapid evolution and development of theoretically sound, robust and efficient finite element methods for the simulation of nonlinear structural dynamics have advanced computer aided vehicle design to the point where the results are trusted with a reasonable degree of confidence.

The advance in super computing is revolutionizing the way vehicles are designed. The success of the finite element simulation to augment or even replace physical prototype in vehicle safety design has accelerated the development of simulation-based optimization in recent years. Early design optimizations of crashworthiness were focused on the component level optimization [1] and occupant safety [2] or airbag related parameter identification [3]. Recently, Yang et al. [4] developed a nonlinear response surface based safety optimization and robustness process, which has been successfully applied to the full vehicle safety design. Sobieski et al. conducted the multidisciplinary design optimization (MDO) for a car body structure under constraints of NVH and roof crush [5]. Yang, Kodiyalam *et al.* included more impact modes with high performance computing [6,7]. This study further extended the previous works to address the uncertainty issue by solving a reliability-based multidisciplinary design optimization problem.

3. RELIABILITY-BASED DESIGN OPTIMIZATION (RBDO)

The Reliability-Based Design Optimization problem can be generally formulated as:

$$\text{Minimize} \quad f(\mathbf{d})$$

$$\text{Subject to} \quad P\left(G_i(\mathbf{X}) > 0\right) > R_i, \ i = 1, \ 2, \cdots, \ N$$

$$\mathbf{d}^l \leq \mathbf{d} \leq \mathbf{d}^u$$

where f and G_i are the objective and constraint functions, respectively, $\mathbf{X}$ is the random design vector, $\mathbf{d}$ is the mean of $\mathbf{X}$, N is the number of probabilistic constraints, ND is the number of design parameters, NR is the number of random parameters, R_i is the desired reliability and the probabilistic constraints are described by the performance function $G_i(\mathbf{X})$ with $G_i(\mathbf{X}) < 0$ indicates failure. It can be rewritten as:

$$\text{Minimize} \ f(\mathbf{d})$$

$$\text{Subject to} \ P\left(G_i(\mathbf{X}) \leq 0\right) - \Phi\left(-\beta_{t_i}\right) \leq 0, \ i = 1, .., N \tag{1}$$

$$\mathbf{d}^l \leq \mathbf{d} \leq \mathbf{d}^u$$

where Φ is the cumulative distribution function for standard normal distribution and β_{ti} is the prescribed target reliability for i^{th} constraint.

The probability of failure is statistically defined by a cumulative distribution function $F_{G_i}(0)$ as

$$P\left(G_i(\mathbf{X}) \leq 0\right) = F_{G_i}(0)$$

$$= \int_{G_i(\mathbf{X}) \leq 0} \cdots \int f_{\mathbf{X}}(\mathbf{X}) d\mathbf{X} \leq \Phi\left(-\beta_{t_i}\right) \tag{2}$$

where $f_x(\mathbf{X})$ is a joint probability density function, which needs to be integrated. To integrate Eq. 2, a dependent, random vector $\mathbf{X}$ is first transformed into an independent, standard normal vector $\mathbf{u}$ through Rosenblatt transformation. In $\mathbf{u}$-space, the most probable point for failure is found by locating the minimum distance between origin and the limit-state or constraint function. The minimum distance is defined as β. Approximate probability integration methods, e.g. the first-order reliability method (FORM), have been widely used to provide efficient and adequately accurate solutions.

Through inverse transformation, the probabilistic constraint in Eq. 2 can be further expressed in two different forms as:

$$\beta_{s_i} = -\Phi^{-1}(F_{G_i}(0)) \geq \beta_{t_i} \tag{3}$$

$$G_{p_i} = F_{G_i}^{-1}\left(\Phi\left(-\beta_{t_i}\right)\right) \geq 0 \tag{4}$$

where β_{s_i} and G_{p_i} are the achieved safety reliability index and the achieved probabilistic performance measure for the ith probabilistic constraint, respectively. The reliability index approach (RIA) uses the reliability index (Eq. 3) to describe the probabilistic constraint in Eq. 1.

C606/026/2002 © IMechE 2002

Similarly, it is referred as the performance measure approach (PMA), if the probabilistic performance measure in Eq. 4 replaces the probabilistic constraint in Eq. 1.

3.1. Reliability index approach (RIA)

In RIA, the first-order safety reliability index β_s is obtained using FORM. It is formulated as an optimization problem, with an implicit equality constraint in a standard $\boldsymbol{u}$-space defined as the limit state function:

$$\begin{aligned} &\text{Minimize} \quad \|\mathbf{u}\| \\ &\text{Subject to} \quad G(\mathbf{u}) = 0 \end{aligned} \tag{5}$$

The optimum point on the failure surface is referred as the most probable failure point (MPFP) $\mathbf{u}^*_{G(\mathbf{u})=0}$. For normal random variables, the Hasofer-Lind (H-L) method [8] can be used to solve Eq. 5. The H-L reliability index can be related to the failure probability if all the variables are statistically independent and normally distributed. For any other situation, Rackwitz and Fiessler (1978) [9] proposed a two-parameter equivalent normal transformation method and Chen and Lind (1983) [10] proposed an extension of the Rackwitz and Fiessler algorithm by using a three-parameter approximation. In this study, the HL-RF (Hasofer, Lind, Rackwitz and Fiessler) is used to perform the reliability analysis in RIA.

3.2. Performance measure approach (PMA)

The first-order reliability analysis in PMA can be formulated as the inverse problem of the first-order reliability analysis in RIA. The first-order probabilistic performance measure G_p is obtained from a nonlinear optimization problem with an n-dimensional explicit sphere constraint in the u-space, defined as:

$$\begin{aligned} &\text{Minimize} \quad G(\mathbf{u}) \\ &\text{Subject to} \quad \|\mathbf{u}\| = \beta_t \end{aligned} \tag{6}$$

The optimum point on a target reliability surface is identified as the most probable point (MPP) $\mathbf{u}^*_{\beta=\beta_t}$. An advanced mean value (AMV) method [11] can be used to solve the inverse PMA problem numerically. However, it is found that the AMV method exhibits poor behavior for concave constraint functions, although it is effective for convex constraint functions. To overcome the difficulties, a conjugate mean value (CMV) method is proposed for treating the concave constraint function. A hybrid mean value method (HMV), which takes advantages of both the CMV and AMV methods, was also presented. It was shown that the HMV method was very robust and efficient for both convex and concave functions [12]. Similar PMA was found in [13,14], however, the hybrid strategy was not used.

There are two major advantages in PMA as opposed to RIA: The PMA is inherently robust and more efficient when the probabilistic constraint is either largely inactive or violated. The reason is that it is easier to minimize a complicated objective function subject to a simple constraint function with known distance (i.e., reliability index) than to minimize a simple objective function subject to a complicated constraint function. PMA always yields a solution, whereas RIA may not produce solutions for certain types of distributions, such as Gumbel or uniform distributions.

4. VEHICLE MODELS AND DESIGN TARGETS

Vehicle safety design is one of the major attributes in car product development. The vehicle structure must be designed to absorb crash energy through structural deformation and attenuate the impact force to tolerable levels when impact occurs. In the real world, all crash modes need to be considered for crash analysis. In this study, four crash modes are considered: full front impact, 50% front offset impact, roof crush, and side impact. The explicit finite element dynamic software RADIOSS was used for all crash simulations.

4.1. Full frontal impact

A full front car crash finite element model is shown in Figure 1. The model contains about 100,000 elements. It crashes into a rigid 90 degree fixed barrier with the speed of 35 MPH. The key safety performance measures include occupant Head Injury Criteria (HIC) and Chest G. Full frontal impact is commonly used to design and validate the vehicle front structure. Federal Motor Vehicle Safety Standards 208 (FMVSS) specifies the safety regulations and test configuration. The design targets for the full frontal impact are based on the FMVSS 208 with a small modification; the occupant HIC value must be less than 370 and the chest G must be less than 42. Note that the numbers may not be realistic, as they are solely used for demonstration of the methodology. Another constraint imposed is the New Car Assessment Program (NCAP) star rating criteria. The occupant NCAP injury probability criteria is derived from the combination of occupant HIC and Chest G. The total occupant probability of severe injury is given by:

$$P_{total} = 1 - (1 - P_{head})(1 - P_{chest}), \text{ where}$$

$$P_{head} = \frac{1}{1 + e^{(5.02 - 0.0035 HIC)}}, \quad P_{chest} = \frac{1}{1 + e^{(5.55 - 0.0693 ChestG)}}, \quad HIC = \left(\left[\frac{1}{t_2 - t_1} \int_{t_1}^{t_2} a \, dt \right]^{2.5} (t_2 - t_1) \right)_{max}$$

where a is deceleration, t_1, t_2 is expressed in seconds and measured during impact, and $(t_2 - t_1)$ is within 36ms. If occupant P_{total} is less than 10%, it receives five stars in NCAP star rating system. The design variables for the front impact are the thicknesses of the subframe and front rail (Table 1).

4.2. 50% Frontal offset impact

A 50% frontal offset impact mode was also considered in the optimization process. The vehicle finite element model is the same as the front impact model. The only difference is the barrier. In the current model, the vehicle crashes into a 90 degree fixed rigid wall with 50% offset (Figure 2). The impact velocity is 40 MPH. The key output from the frontal offset impact is the toeboard intrusion. The design target for toeboard intrusion is set to be less than 11 inches. The design variables used for 50% offset front impact is same as those used for full front impact (Table 1).

4.3. Roof crush

Vehicle roof crush is a federally mandated requirement intended to enhance passenger protection during a rollover event. The test procedure is defined in FMVSS 216. The finite element roof crush model for this study is converted from a NVH model, as shown in Figure 3. The unnecessary parts in the NVH model are deleted and some missing parts are added in the roof crush model, e.g., very detailed side doors were added and the glasses are refined. The total

C606/026/2002 © IMechE 2002

number of elements for roof crush is about 120,000. A 72 inches by 30 inches rectangular ram is added to perform the roof crush as specified by the FMVSS 216. The longitudinal axis of the ram (see Figure 3) is at a forward angle (side view) of 5 degrees below the horizontal, and is parallel to the vertical plane through the vehicle's longitudinal centerline. The lateral axis is at a lateral outboard angle, in the front view projection, of 25 degrees below the horizontal. The lower surface is tangent to the surface of the vehicle and initial contact point is on the longitudinal centerline of the lower surface of the ram and 10 inches from the forward most point of the centerline. In roof crush simulation, the ram normal speed is set as 7.5 MPH.

As described in the FMVSS 216, the force generated by vehicle resistance must be greater than 5,000 lbs (22,240 N) or 1.5 times the vehicle weight, which ever is less, through 5 inches of ram displacement. In this study, the roof crush resistant force was set to be 6,000 lbs (≈27 kN). The door thickness and material yield stresses are chosen as the design variables (Table 1).

4.4. Side Impact

For side impact protection, the vehicle design should meet the requirements for the National Highway Traffic Safety Administration (NHTSA) side impact procedure (FMVSS 214) or European Enhanced Vehicle-Safety Committee (EEVC) side impact procedure. In our study, the EEVC side impact test configuration is used. The dummy performance is the main concern in side impact, which includes head injury criterion (HIC), chest V*C's (viscous criterion) and rib deflections (upper, middle and lower). These dummy responses must at least meet EEVC requirements. The finite element vehicle model along with moving deformable barrier model is shown in Figure 4. A finite element dummy model is also employed for prediction. Other concerns in side impact design are the velocity of B-Pillar at middle point and the velocity of front door at B-Pillar. The total number of elements in this model is about 100,000.

The moving deformable barrier position is defined in the EEVC side impact procedure. All nodes of the moving barrier are assigned an initial velocity equal to 50 km/h. For side impact, the increase of gage design variables tends to a get better dummy performance. However, it also increases vehicle weight, which is undesirable. Therefore, a balance must be sought between weight reduction and safety concerns. The objective is to reduce the weight while imposing safety constraints on the dummy. The dummy safety performance is usually measured by EEVC side impact safety rating score. In the EEVC side impact safety rating system, the safety rating score depends on four measurements of the dummy: HIC, abdomen load, rib deflection or V*C, and pubic symphysis force. In this study, the dummy chest V*C and rib deflection are used to measure the safety performance.

5. NUMERICAL OPTIMIZATION

The optimization problem is to minimize the vehicle weight subject to design constraints imposed on roof crush, full front impact, 50% frontal offset impact, and side impact. The deterministic problem and reliability-based design optimization problem can be formulated as follows:

	Deterministic Problem	**Reliability-Based Problem**
Minimize	Vehicle Weight	Vehicle Weight
Subject to	Roof Crush Constraints:	Roof Crush Constraints:
	Crush distance D ≤ 5"	P(Crush distance D ≤ 5") ≥ 90%
	Critical peak load P_{cr} ≥ 27kN	P(Critical load peak P_{cr} ≥ 27kN) ≥ 90%
	Full Frontal Impact Constraints:	Full Frontal Impact Constraints:
	HIC ≤ 370	P(HIC ≤ 370) ≥ 90%
	Chest G ≤ 42	P(Chest G ≤ 42) ≥ 90%
	P_{total} ≤ 10 %	$P(P_{total}$ ≤ 10 %) ≥90 %
	50% Frontal Offset Impact Constraints:	50% Frontal Offset Impact Constraints:
	Toe board intrusion ≤ 11"	P(Toe board intrusion ≤ 11") ≥90%
	Side Impact Constraints:	Side Impact Constraints:
	V*C ≤ 0.58	P(V*C ≤ 0.58) ≥ 90%
	$D_{upper\ rib,\ middle\ rib,\ lower\ rib}$ ≤ 27.2	$P(D_{upper\ rib,\ middle\ rib,\ lower\ rib}$ ≤ 27.2) ≥ 90%

Figure 5 shows the MDO data flow. The optimization problem includes 10 system or global design variables, which are common to all crash modes. The local design variables which are important in the front impact (5), front offset impact (5), roof crush (10), and side impact (5) are also identified. All the design variables are treated as random variables with normal distribution, as shown in Table 1. It is noted that the local design variables for the frontal impact and frontal offset impact are the same, namely there are 30 design variables totally in this problem.

The objective and constraint functions are approximated by using the Response Surface Method [15]. The uniform Latin Hypercube Sampling [16] and Stepwise Regression methods were used for selecting the initial design points in the exploratory design space and construct the response surfaces, respectively. The second order polynomials for the regression are used to construct the nonlinear response surfaces for front impact, side impact and frontal offset impact. The weight and resistant force in the roof crush are approximated using linear functions.

6. RESULTS AND DISCUSSIONS

Both PMA and RIA are used to perform the reliability analysis and the sequential quadratic programming method is for the design optimization to solve a full vehicle MDO problem with probability constraints imposed on four impact modes, namely, frontal impact, roof crush, frontal offset impact and side impact. After optimization, the Monte Carlo (MC) method is employed to validate the RBDO results by targeting the mean values at the optimum design with predetermined standard deviations.

Conventional deterministic optimization is first performed to find a deterministic multidisciplinary optimum design, which is compared with that of the RBDO results. Table 2 summarizes both deterministic optimization and RBDO results. In the deterministic optimization, a feasible design is obtained and the vehicle weight is reduced by 16.8kg. The reliability assessment analysis using MC method with 10,000 sampling points is performed on the deterministic optimum design, the result shows that the reliability of the 2nd intrusion is only 62.1% while the remaining constraints are 100%.

To increase the reliability of the constraints, the RBDO process is performed using both PMA and RIA. The Hybrid Mean-Value (HMV) that combines the Advanced Mean-Value (AMV) and Conjugate Mean-Value (CMV) methods is used for the PMA. The method first identifies whether the performance function is convex or concave and then chooses different numerical algorithms adaptively for the MPP search. As for the RIA, the HL-RF (Hasofer, Lind, Rackwitz, and Fiessler) method is used. The HL-RF method consists of a specific iterative scheme that is simple and exhibits rapid convergence for most problems.

The reliability of each constraint is targeted at 90% in this research. It shows that both methods achieve a feasible optimum design. The designs from both methods are similar for most design variables, as shown in Table 1. The different design variables are highlighted. The vehicle weight using RIA and PMA is reduced by 13.1kg and 13.8kg, respectively. The reliability assessment analyses using 10,000 MC sampling points show that the reliability of the 2nd intrusion for RIA and PMA is increased from 62.1% to 91.2% and 90.7%, respectively, while all other constraints remain in 100%. It is noted that the number of function and sensitivity evaluations for PMA is only 660, while RIA requires 6258.

7. CONCLUSIONS

This paper has successfully demonstrated the feasibility of applying the reliability-based design optimization method to achieve a robust crashworthiness design of a full vehicle in multi-crash scenarios. Four impact modes: frontal impact, roof crush, frontal offset impact, and side impact are studied in this research. Both reliability index approach and performance measure approach are applied and feasible designs are obtained. The performance measure approach, however, is more robust and efficient. It requires much less function and sensitivity evaluations compared to the reliability index approach.

8. ACKNOWLEDGMENTS

The authors acknowledge the support of S. Kodiyalam of SGI for the assistance and the dedicated time on SGI Origin 3000 machines, T. Tyan of Ford Motor Company for providing the vehicle crash models and consultation, and RADIOSS Consulting Corporation for providing RADIOSS software license for crash simulations. Iowa team has been supported by the Automotive Research Center sponsored by the U.S. Army TARDEC.

9. REFERENCES

1. R. J. Yang, L. Tseng, L. Nagy, and J. Cheng, "Feasibility Study of Crash Optimization", ASME, Vol. 69-2, pp. 549-556, 1994.
2. L. F. P. Etman, J. M. T. A. Adrisens, M. T. P. Von Slagmaat, and A. J. G. Schoofs, "Crashworthiness Design Optimization using Multipoint Sequential Linear Programming," Structrual Optimization, Vol 12, pp. 222-228, 1996.
3. N. Stander, "Optimization of nonlinear Dynamic Problems using Successive Linear Approximations, AIAA-2000-4798.

4. R. J. Yang, L. Gu, L. Liaw, C. Gearhart, C.H. Tho, X. Liu, and B. P. Wang, "Approximations For Safety Optimization Of Large Systems," Proceedings of ASME Design Engineering Technical Conferences, September, Baltimore, Maryland, 2000.

5. J. Sobieski, S. Kodiyalam, R.J. Yang, "Optimization of Car Body under Constraints of Noise, Vibration, and Harshness (NVH), and Crash," Structural and Multidisciplinary Optimization, Vol. 22, No. 4, 295-306, 2001.

6. R. J. Yang, L. Gu, C.H. Tho, J. Sobieski, "Multidisciplinary Design Optimization of A Full Vehicle with High Performance Computing," AIAA-2001-1273, In Proceedings of 42nd AIAA SDM Conference, Seattle, Washington, 2001.

7. S. Kodiyalam, R. J. Yang, L. Gu, and C.H. Tho, "Large-Scale, Multidisciplinary Optimization of a Vehicle System in a Scalable, High Performance Computing Environment," DETC2001/DAC-21082, in Proceedings of ASME Design Engineering Technical Conferences, Pittsburgh, Pennsylvania, September 2001.

8. A. M. Hasofer and N. C. Lind, "Exact and Invariant Second Moment Code Format." *Journal of the Engineering Mechanics Division*, ASCE. Vol. 100, p 111-121, 1974.

9. R. Rackwitz, and B. Fiessler, "Structural Reliability under Combined Random Load Sequences," Computers and Structurals. Vol. 9, p484-494, 1978.

10. X. Chen and N. C. Lind, "Fast Probability Integration by Three Parameter Normal Tail Approximation," Structural Safety. Vol. 1, p 269-276, 1983.

11. J. Tu and K. K. Choi, "A New Study on Reliability Based Design Optimization," *ASME Journal of Mechanical Design*, Vol. 121, No. 4, 1999, pp. 557-564.

12. K. K. Choi and B. D. Youn, "Hybrid Analysis Method for Reliability-Based Design Optimization," DETC2001/DAC-21044, in Proceedings of ASME Design Engineering Technical Conferences, Pittsburgh, Pennsylvania, September 2001.

13. J. Wu, Y. Shin, R. Sues, and M. Cesare, "Safety-Factor Based Approach for Probability-Based Design Optimization, AIAA-2001-1522, In Proceedings of 42nd AIAA SDM Conference, Seattle, Washington, 2001.

14. X. Du and W. Chen, "A Most Probable Point-Based Method for Efficient Uncertainty Analysis," Design Manufacturing, Vol. 4, No. 1, pp. 47-66, 2001.

15. R. J. Yang, N. Wang, C.H. Tho, and J. P. Bobineau, "Metamodeling Development for Vehicle Frontal Impact Simulation," DETC2001/DAC-21012, In Proceedings of ASME Design Engineering Technical Conferences, Pittsburgh, Pennsylvania, September 2001.

16. K. Fang, D. Lin, P. Winker, and Y. Zohang, "Uniform Design: Theory and Application." Technomettrics, Vol. 42, pp 237-248, 2000.

Table 1. MDO random variables

No	Design Variable	Initial Design	Lower Bound	Upper Bound	Standard Deviation	Det. Opt	RBDO	
							RIA	PMA
Common Design Variables (10)								
1	Windshield (mm)	3.8	2.6	5.0	0.114	2.6	**2.74**	**2.6**
2	Roof panel (mm)	0.7	0.6	1.5	0.021	0.6	0.63	0.6
3	Roof rail (mm)	1.0	0.6	1.5	0.030	0.6	0.61	0.6
4	Front Roof cross member (mm)	1.0	0.6	1.5	0.030	0.6	0.61	0.6
5	Rear roof cross member (mm)	0.9	0.6	1.5	0.027	0.6	0.6	0.6
6	A-Pillar (mm)	0.8	0.6	1.5	0.024	0.803	**0.86**	**0.8**

#	Variable							
7	B-Pillar 1 (mm)	1.0	0.6	1.5	0.030	0.6	**0.67**	**0.6**
8	B-Pillar 2 (mm)	0.8	0.6	1.5	0.024	0.6	0.6	0.6
9	B-Pillar 3 (mm)	1.35	1.0	2.0	0.040	1.0	**1.13**	**1.0**
10	C-Pillar (mm)	0.8	0.6	1.5	0.024	0.6	0.61	0.6
	Roof Crush Design Variables (10)							
11	Front door (mm)	0.7	0.4	1.0	0.021	0.7	0.7	0.7
12	Front door inner (mm)	0.7	0.4	1.0	0.021	0.7	**0.4**	**0.7**
13	Rear door (mm)	1.0	0.7	1.3	0.030	1.0	0.98	1.0
14	Mat A-Pillar 1 (GPa)	0.207	0.192	0.345	0.006	0.207	**0.192**	**0.207**
15	Mat A-Pillar 2 (GPa)	0.207	0.192	0.345	0.006	0.207	0.207	0.207
16	Mat A-Pillar 3 (GPa)	0.207	0.192	0.345	0.006	0.207	0.207	0.207
17	Mat B-Pillar 1 (GPa)	0.207	0.192	0.345	0.006	0.207	0.207	0.207
18	Mat B-Pillar 2 (GPa)	0.207	0.192	0.345	0.006	0.207	0.205	0.207
19	Mat Front door inner 1 (GPa)	0.207	0.192	0.345	0.006	0.207	0.207	0.207
20	Mat Front door inner 2 (GPa)	0.207	0.192	0.345	0.006	0.207	**0.192**	**0.207**
	Frontal Crash and 50% Frontal Offset Crash Design Variables (5)							
21	Subframe (mm)	2.0	1.0	3.0	0.060	1.0	1.0	1.0
22	Rail 1 (mm)	1.9	1.0	3.0	0.057	1.0	1.0	1.0
23	Rail 2 (mm)	1.9	1.0	3.0	0.057	1.0	**1.1**	**1.0**
24	Rail 3 (mm)	1.9	1.0	3.0	0.057	1.0	**1.4**	**1.0**
25	Rail 4 (mm)	2.4	1.0	3.0	0.072	2.31	**1.2**	**1.0**
	Side Impact Design Variables (5)							
26	Door Reinf. (mm)	1.6	1.2	2.0	0.048	1.5	**1.2**	**1.6**
27	Rocker Outer (mm)	1.1	0.7	1.5	0.033	1.1	**0.77**	**1.1**
28	Rocker Inner (mm)	1.75	1.2	2.3	0.0525	1.74	**1.2**	**1.75**
29	Cross Member 1 (mm)	2.3	1.5	3.0	0.069	2.3	**2.3**	**2.4**
30	B-Pillar Reinf. (mm)	2.2	1.5	3.0	0.066	2.26	**3.0**	**2.2**

Table 2. Deterministic optimization and RBDO results

Sub-system	Attribute	Baseline	Target	Det. Opt.	RBDO	
					RIA	PMA
Frontal Crash	HIC	366	P($\leq$ 370)$\geq$90%	298.8	297.3	298.8
	Chest G (g)	39.9	P($\leq$ 42) $\geq$90%	41.4	40.9	41.4
	P_{total} (%)	8.0	P($\leq$ 10) $\geq$90%	8.1	7.9	8.1
Roof Crush	Resistant force (kN)	34.7	P($\geq$ 27) $\geq$90%	42.2	39.5	42.2
	Intr 1 (in.)	9.9	P($\leq$ 11) $\geq$90%	10.0	10.4	10.5
				9.8	10.4	10.4
Offset	Intr 2 (in.)	10.9	P($\leq$ 11) $\geq$90%	*(62.1%)*	*(91.2%)*	*(90.7%)*

Crash	Intr 3 (in.)	10.5	P($\leq$ 11) $\geq$90%	9.7	10.5	10.5
	Intr 4 (in.)	9.7	P($\leq$ 11) $\geq$90%	9.7	9.5	9.7
	Intr 5 (in.)	10.3	P($\leq$ 11) $\geq$90%	9.9	10.3	10.4
Side	Disp 1 (mm)	23.1	P($\leq$ 27.2) $\geq$90%	23.3	22.8	23.3
	Disp 2 (mm)	26.0	P($\leq$ 27.2) $\geq$90%	26.0	24.9	26.0
Impact	Disp 3 (mm)	26.9	P($\leq$ 27.2) $\geq$90%	26.8	25.9	26.8
	V*C 1	0.48	P($\leq$ 0.58) $\geq$90%	0.47	0.46	0.47
	V*C 2	0.53	P($\leq$ 0.58) $\geq$90%	0.53	0.50	0.53
	V*C 3	0.55	P($\leq$ 0.58) $\geq$90%	0.54	0.51	0.55
System	**Weight (Kg)**	**1740.5**	**Minimize**	**1723.7**	**1727.4**	**1726.7**

Note: Total number of function and sensitivity evaluations: **6258** (RIA), **660** (PMA)

Figure 1. Frontal impact model

Figure 3 Roof crush model

Figure 2. 50% Frontal offset impact model

Figure 4. Side impact model

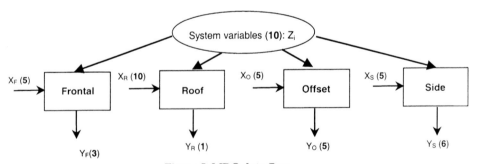

Figure 5. MDO data flow

C606/026/2002 © IMechE 2002

C606/002/2002

An analysis methodology for the simulation of vehicle pedestrian accidents

N LE GLATIN, M V BLUNDELL, and **G BLOUNT**
School of Engineering, Coventry University, UK

ABSTRACT

This paper outlines an analysis methodology for the validation of pedestrian accident simulations to help engineers and designers to apply simulation is the area with increasing confidence. The research undertaken will support the automotive industry and the coming requirement to design future vehicles that offer improved protection to the pedestrian on impact. The simulations performed aim to represent more realistic pedestrian accidents, as they occur in the real world. The work presents particular problems in terms of validation of the individual components of the system. In particular the modelling of the pedestrian presents challenges in terms of establishing the biofidelity of the model. To facilitate the analysis of real world pedestrian to vehicle accidents, a Design of Experiment (DoE) methodology has been adopted to manage a wide range of accident scenarios. The DoE statistical model has been used to provide conclusions on predictive pedestrian post-impact kinematics and injury criteria.

1 - INTRODUCTION

For more than twenty years, the automotive industry has made continuous progress in the development of safety devices that provide drivers and passengers with high levels of protection during a vehicle crash. In combination with the development of these mechanical systems, analysis methodologies and engineering computer based design tools have also been developed to design the systems. This, in the main, has involved the development of finite element and multibody systems analysis programs to model vehicles, dummies, airbags, and restraint systems.

As we move forward into the 21st century, the automotive industry is turning its attention to the exterior of the vehicle and the design of "pedestrian friendly" vehicles in order to respond to developments in legislation.

Over the ten years period identified in (1), 46,000 people have been killed on UK roads, with another 580,000 seriously injured. This does not take into account unreported accidents, which some estimates put at half as many again. On this basis, there have been an estimated 900,000 people killed and injured on UK roads in the past ten years.

As a result the European Experimental Vehicles Committee (EEVC) formed a working group to assess and develop test methods for evaluating pedestrian protection for passenger cars at the beginning of the 1990's.

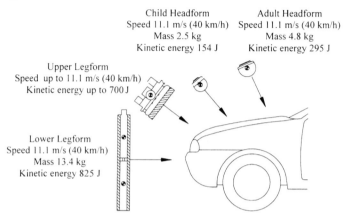

Figure 1: EEVC WG17 Proposed Pedestrian Impactor Test Regulation

The current proposed pedestrian protection regulations based on the findings of the EEVC WG17 (2) propose different test methods, shown in Figure 1, and criteria that should enable a pedestrian to survive a frontal impact without death or serious injury, with a car travelling at 40 km/h.

Current practice for designing vehicles aims to use computer programs to simulate test procedures used within the laboratory. Further studies have been established which extend simulations beyond the procedure outlined above to investigate pedestrian impact events under real "world" conditions (3-4), by developing more detailed biofidelic pedestrian models and vehicle models.

The work at Coventry University, in collaboration with the Motor Industry Research Association (MIRA) focuses on the application of Design of Experiment (DoE) analysis techniques, combined with Finite Element (FE) methods and Multibody System (MBS) analysis techniques, to recreate simulations of vehicle to pedestrian impacts scenarios.

The analysis methodology presented in the paper has been subdivided into three individual stages:

- Validation of a generic FE/MBS vehicle model, by correlating results with EEVC WG10 impactor physical models.

- Application of a pedestrian human body model. The resultant kinematics from the interaction between a Madymo pedestrian human body model and a FE/MBS vehicle model (see Figure 5) have been compared with published results from impacts tests with human cadavers (4).

C606/002/2002 © IMechE 2002

- Evaluation of the pedestrian vehicle system using mathematical pedestrian thrown distance models (5). These models relate the vehicle velocity at impact, with the post-impact pedestrian thrown distances, based on data collected from real world accidents.

Following this methodology, the developed simulation tools have been validated on three different levels to compensate for the impossibility of recording data during a real world accident.

An important aspect of this work is the application of Design of Experiment analysis techniques, to represent a wide range of vehicle dynamics, pedestrian kinematics and impact configurations. The results and experience gained from this study will enable a better understanding of the effects of vehicle dynamics and pedestrian kinematics on the overall injury criteria and post-impact kinematics of the pedestrian.

2 – SIMULATION OF REAL WORLD PEDESTRIAN VEHICLE ACCIDENTS

2.1 - Validation of a generic FE/MBS vehicle model

This section focuses on the development of a FE/MBS vehicle model used in the pedestrian to vehicle accident simulations. The front end of the vehicle model is made up of five FE sub-models, as shown in Figure 4, in contrast to the rear of the vehicle, which is modelled using the MBS rigid body analysis technique. Particular attention has also been paid to the design of vehicle components that would interact with the pedestrian during an impact event.

The objective of this phase of work was to develop a FE vehicle model, which would produce realistic interaction between the impactor models and the front end of the vehicle structure, in order to gain sufficient confidence in the use of the model for further pedestrian/vehicle simulation studies. This has been achieved by recreating impact scenarios using validated MADYMO EEVC WG10 impactor models available for this study(6).

A parallel correlation exercise was also completed to compare output results (Lower Leg acceleration, Knee bending angle, Femur shear force) of physical pedestrian impactor testing with the MADYMO EEVC impactor simulation models.

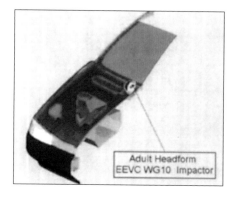

Adult Headform
EEVC WG10 Impactor

Figure 2: Generic FE/MBS Front End Vehicle Model

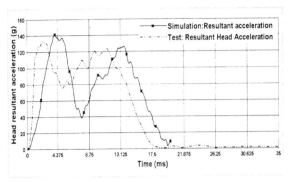

Figure 3: Adult Head Impactor Res. Acceleration (g)

The results of this study indicate that the impactor computer models demonstrate a good correlation with physical impactors designed to recreate a characteristic pedestrian accident.

Figure 3 shows, an example of the resultant acceleration of the adult headform, impacting the centre line of the bonnet with a secondary impact with the front panel of the vehicle, see Figure 2.

2.2 – Application of a pedestrian human body model

As noted in the introduction, an inherent constraint in this study is the lack of access to real world pedestrian accident data to validate pedestrian to vehicle impact simulations. To overcome this issue it was decided to evaluate the response of each individual component of the pedestrian vehicle system. This section discusses evaluation of a multibody MADYMO Pedestrian human body model (See opposite) developed by TNO (6)

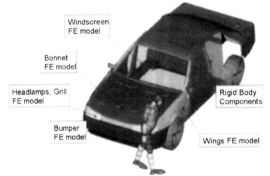

Figure 4: Pedestrian to Vehicle
Accident simulation

The validity of the pedestrian human body model was evaluated using published results from impact tests with cadavers (4).

The responses of the pedestrian model were reproduced for a set of cadaver test configurations. Five tests have been identified to represent a wide of range different vehicle stiffness and geometry characteristics.

Each simulation was set up with the pedestrian positioned in a walking posture and the knee extended, balanced in an upright position. Post impact trajectories of the pedestrian model were measured at four different locations on the human body; Head CG, Pelvis, Knee and Foot.

Table 1: Test set up of cadaver tests

Test	Vehicle velocity	Bumper Level	Bumper lead	Hood edge level	Age/Sex	Weight	Height
T1	25 km/h	0.38 m	0.06 m	0.73 m	54/Male	75 kg	1.80 m
Pedestrian Human Body Model					18-70/Male	88 kg	1.75 m

C606/002/2002 © IMechE 2002

A visual comparison of the resultant kinematics between a human cadaver test, outlined for just one of the five tests in Table 1 and the equivalent computer simulation is presented below. It appears that the overall kinematics of the pedestrian, shown in Figure 5, is in good agreement with recorded high-speed film from impact tests.

MADYMO computer simulation

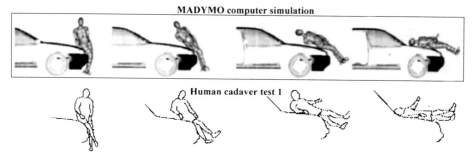

Figure 5: Visual comparison between human cadaver test 1 and computer simulation model starting at 50 ms after first contact, Time Step Δt=50 ms

2.3 – Validation of the pedestrian/vehicle system model

Following the validation of the FE vehicle model and the application of a pedestrian human body model, a method of validating the entire pedestrian to vehicle impact event was investigated. To establish confidence in the use of the simulation tool described here, predicted pedestrian throw distances were calculated using mathematical models (5). Developed from real world vehicle to pedestrian accident data collection. Using regression analysis techniques, the authors have developed a set of equations to calculate the post-impact thrown distances based on vehicle velocities at impact.

An example of the validation of the vehicle/pedestrian system is presented in Figure 6. The accident scenario has been extracted from the simulation matrix generated by the Design Of Experiment analysis study presented in the next section.

	Mathematical model * Eq (1)	Computer simulation
Pedestrian Throw Distance	19.5 m	18.6 m

* Eq (1): $Vv = 11.4(dt)^{1/2} - 0.4$

Vv: Vehicle velocity at impact (50 km/h)
dt : Pedestrian Thrown distance

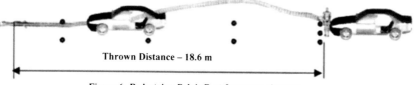

Thrown Distance – 18.6 m

Figure 6: Pedestrian Pelvis Post-Impact trajectory

3 – DESIGN OF EXPERIMENTS (DoE)

This section discusses the application of DoE techniques to analyse the results from real world pedestrian to vehicle accident simulations. The objective of the study was to carry out an analysis of the main variables influencing the post impact kinematics and the location and severity of injuries sustained by the pedestrian when impacted with a vehicle.

3.1 – DoE techniques theory

It should be reiterated that two types of models have been utilised in this work; a FE/MBS simulation model (more familiar to engineers) and a DoE statistical model used to manage an extensive range of simulations.

A DoE model contains several features:

- *Outputs and responses* – These are the responses required from the system under investigation, examples being Knee bending angle, Head Injury Criteria (HIC) and upper leg loads.
- *Variables* – These are the major factors of interest that will affect the outputs and include for example, the vehicle dynamics, initial pedestrian position and velocity, impact location and bone fractures.

The different parameters that represent the system interact in a complex manner and the system studied is non-linear. These factors contribute to difficulties in the interpretation of overall system behaviour. If the system is modelled by critical variables then some formal method for 'mapping out' the behaviour of the system becomes a necessity.

A Design of Experiments analysis technique is a carefully established series of trials, based on sound mathematical theory, where the system is "interrogated" for set combinations of the variables. This is a highly cost-effective technique for parametric studies that is used to build an accurate empirical model of the system. The DoE model allows the prediction of various functions of the responses within well-defined boundaries of probability, by relating the responses to the variables, both individually and in combination.

The key insight is that building an accurate statistical model of the complex multi-variable system produces a number of important benefits that would not be available in any other way, and for a sensible cost regarding the number of simulation runs carried out. Once a statistical model has been designed, it can be used to investigate and analyse the system in a number of useful ways, including:

- *Visualisation* – The statistical model can be used to generate a set of 3-D 'slices', as shown in Figure 7, to represent the behaviour of the system with respect to two variables, with other variables fixed constant.

- *Optimisation* – The DoE statistical model can be used to calculate changes quickly. This allows an optimisation of the system via the statistical model that could not realistically be attempted directly by FE and MBS simulation analysis techniques.

C606/002/2002 © IMechE 2002

A Design of Experiments analysis is a three-stage process; experimental design, statistical data modelling and application/analysis of the statistical model.

Practical methods for choosing an experimental design and building accurate multi-dimensional model for analysis, prediction and optimisation were given foundation in the 1950's (7). This period of research and practice produced a set of techniques known as Response Surface Methodology (RSM).

These design and analysis techniques have been used to build the system's responses presented in the next section, as a function of the variables under investigation. Each part of this process has required the use of the commercial DoE software Design-Expert that incorporates the above-mentioned techniques.

3.2 – Experimental Design of the Pedestrian/Vehicle DoE model

Previous experience has shown that planning the selection of the variables and responses is the most important phase of the Design of Experiments. Errors made in the Experimental Design will produce poor results weeks or months later during the analysis phase.

The design phase requires information regarding the variables that are likely to be most influential on the responses defined to interpret and analyse the pedestrian/vehicle system of interest. The range over which the variables should be varied, and the expected output behaviour, also needs to be considered.

The information generated in the design phase helps to create an optimal Experimental Design that defines the variables settings at which the FE simulation should run in matrix form.

There is always a trade off between the number of variables, outputs, number of simulation runs, time and cost. The computer simulation time for each individual pedestrian impact analysis was between 3-8 days. This was for pedestrian to vehicle accident events simulated for up to 3 seconds in order to analyse secondary impacts of the pedestrian on the road.

It was decided with regard to computer resources and availability, to investigate a maximum number of 7 variables, listed in Table 2. For the purpose of this study a D-Optimal cubic design has been used to generate a design matrix of 130 simulation runs.

DoE variables		Range	
		Min	Max
V1: Vehicle velocity at impact		15 km/h	50 km/h
V2: Pedestrian velocity at impact		0 km/h	7 km/h
V3: Pedestrian lateral position		-0.86 m	0.86 m
V4: Pedestrian walking posture		- 23 deg	23 deg
V5: Normal vehicle acceleration		-0.6 g	+ 0.5g
V6: Vehicle braking time after To		0 sec	1 sec
V7: Pedestrian leg Fracture	Upper leg	5 kN	7 kN
	Lower leg	3 kN	5kN

Table 2: DOE Variables

To define the various factors that could influence the kinematics and injuries of the pedestrian model a literature review (8) has been carried out, completed with technical support of engineers working in the field of pedestrian safety.

For the purpose of this study the following variables have been defined as described and below:

- V5: Describes the normal vehicle acceleration at the time of impact. Preceding an impact with a pedestrian in an urban area, the driver could be accelerating, decelerating the vehicle ore driving at constant speed.

- V6: Takes into account the time at which an emergency braking deceleration of 0.75g will be applied during the impact event after To and allows a variable delay to be incorporated into the model without the detailed behaviour of driver reaction time models,
- V3: Represents the initial pedestrian standing lateral position relative to the front end of the vehicle,
- V4: Represents the angular position of the pedestrian legs at impact To,
- V7: Takes into account bones (tibia and femur) fracture level.

To represent the pedestrian/ vehicle system under investigation, four outputs were identified from the set of simulation results available. The post-impact trajectory and injuries criteria defined to analyse the interactions between a pedestrian and vehicle are listed below:

- Pedestrian thrown distance,
- Maximum HIC due to an impact with the vehicle and the ground,
- Lower leg bending angle,
- Femur load.

3.3 – Statistical data modelling using regression techniques

The modelling phase takes the sample responses generated by the experiment (i.e., the simulation results) and builds an accurate statistical model of the system studied. The validity of the DoE statistical responses have been evaluated using statistical tools. These include normal probability plot of the studentized residuals to check for normality of residuals, studentized residuals versus predicted values to check for constant error, predicted values versus actual values and Box-Cox plot power transformations

Once the system model has been formed, it can be manipulated mathematically in a number of ways. An example is highlighted in Figure 7, where the influence of variable V1 (Vehicle Velocity in m/s) and V7 (Pedestrian Leg Fracture level) on the knee bending angle response is represented. It is also possible to analyse the set of all different combinations of variables and draw conclusions on predictive post-impact kinematics and injury location and severity.

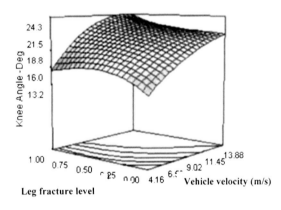

Figure 7: Pedestrian thrown distance

C606/002/2002 © IMechE 2002

3.4 – Validation of the statistical model using point prediction analysis

Once the Doe statistical model has been created using outputs from the 130 simulation runs, the statistical criteria outlined in Section 3.2 can be used to check the validity of the statistical model obtained using the regression analysis techniques.

A follow on exercise from this is to use the MBS/FE model to run random simulations within the design space and compare outputs from these simulations with those predicted using the statistical model. Up to ten FE/MBS simulation runs were carried out to quantify the validity of the DoE statistical model.

3.5 – Statistical data modelling using Gaussian Process techniques

Research beginning in the late 1980's suggested more effective methods than RSM techniques for both choosing an experimental design and building accurate multi-dimensional models for analysis, prediction and optminisation of the underlying system. It has now become apparent that response surface methodology techniques are not ideal. To build justifiable RSM models there should be uncorrelated and normally distributed random error in the outputs to be modelled. This will not usually be the case with outputs from simulations.

The second approach presented in this paper is based on what ACS consultancy calls Gaussian Process (GP) modelling techniques. GP models are derived from a different set of statistical theory from regression models – Stochastic processes, Random fields, Bayesian function approximation, Spatial statistics. Some of the benefits are: they can fit very complex behaviour with relatively few runs; do not have to assume identical independently-distributed (IID) errors in the data.

Figure 8 presents an example of "1D and 2D" trend plots derived from the statistical model used to represent the pedestrian thrown distance response. The 1D trend plots show the behaviour of the response to each input variable, averaged over the six input variables. In the absence of strong interactions between input variables, these trends give a good idea of the importance and nature of the effect of individual input variables.

The 2D trend plots represent the average behaviour for each 2D subspace, average over all combinations of the other five variables.

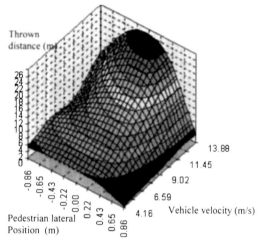

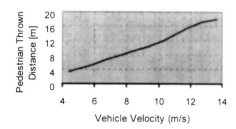

Mean trend (ignoring interactions)

Figure 8: 1D & 2D Trend plots

4 - CONCLUSIONS

This paper outlines an analysis methodology use to simulate real world pedestrian to vehicle accidents. The methodology of validation discussed is a three stages process, which includes the validation of a FE/MBS vehicle model, the application of a pedestrian human body model and finally the validation of the vehicle/pedestrian simulation system using real world pedestrian accident data.

Using the simulation tool that has been created, a DoE analysis study has been performed to simulate a wide range of pedestrian impact events, and predict post-impact kinematics and injuries location and severity sustained by the pedestrian human body model.

ACKNOWLEDGEMENTS

The authors wish to acknowledge Mike Hopkins from ACS Consultancy for his valuable time and assistance with the statistical modelling of the DoE analysis study.

5 - REFERENCES

(1): Lawrence G.J.L., Hardy B.J, " Costs and Benefits of the EEVC Pedestrian Impact Requirements." Project Report 19, TRL, UK, (1993)

(2) EEVC/CEVE (1998), "Improved test methods to evaluate pedestrian protection afforded by passenger cars". Working Group 17 Report, December 1998.

(3) Howard, M.S., Thomas, A.V., Koch. W, (2000), "Validation and Application of a Finite Element Pedestrian Humanoid Model for Use in Pedestrian Accident Simulations".

(4) Ishikawa, H. et al, (1993), "Computer Simulation of Impact Response of the Human Body in Car-Pedestrian accidents", Proceedings of the 37th Stapp Car Crash Conference.

(5) Happer, A., Toor, A., (2000), "Comprehensive Analysis Method for Vehicle/Pedestrian Collisions", SAE 2000 World Congress, Paper 2000-01-0846.

(6) MADYMO (1997). MADYMO Version 5.4 Manuals. TNO Road-Vehicles Institute, Delft, Netherlands.

(7) Box, G., Hunter, W., (1978), "Statistics for Experiment", John Wiley & Sons, Inc.

(8) Le Glatin, N., Blundell M., Thorpe, S., (2000), "Human Body Modelling Techniques for Use with Dynamics Simulation", Multi-body Dynamics: Monitoring Simulation Techniques.

C606/002/2002 © IMechE 2002

C606/027/2002

Sensitivity of transmission noise and vibration to manufacturing and assembly process drift and variability

S M ATHERVALE
Ford Research Laboratory, Dearborn, Michigan, USA
G D GARDNER and **M TRENT**
Powertrain Operations, Ford Motor Company, Livonia, Michigan, USA

ABSTRACT

Low transmission noise and vibration is a key consumer expectation. Besides inherent design influence, manufacturing and assembly process have significant impact on transmission noise and vibration. Manufacturing and assembly contributors like dimensional deviations, deflections, misalignments, can be lumped together in the form of transmission error. In this work, empirical regression equations between transmission error, shaft assembly, bearing clearance, and transmission noise and vibration are developed for the Ravigneaux type planetary gear system. The transmission error sources include lead deviation, pinion hole position and misalignment. A method to predict transmission error for the Ravigneaux type gear system is also presented. The statistical distribution for both noise (sound pressure) and the drive shaft torsional acceleration are determined for a given tolerance and nominal values for transmission error and bearing clearance.

1. INTRODUCTION

Design and manufacture of automotive transmissions entails a delicate balance between key functional requirements and prominent consumer comforts. Examples of key functional requirements can be torque transmissibility or shift-quality, whereas low transmission noise and vibration are prominent consumer expectations. Transmission noise and vibration is inherent to each transmission design. Suppression of the transmission noise and vibration can be accomplished on three levels. First, the noise and vibration source must be reduced. Second, the noise and vibration path to the passenger compartment must be designed to attenuate noise and vibration. Third, active control of noise and vibration can be used. Reduction in the noise and vibration intensity at the source is of interest in this paper. Specifically, the focus is on noise and vibrations due to the planetary gear system.

Transmission noise and vibration related to gears can be attributed to dynamic deflections and dimensional deviations. Although the root-cause of noise and vibration can be identified qualitatively, there is no clear quantitative relationship between the two. Thus, empirical methods must be used to address this issue. A typical empirical approach is to design an experiment with two to three levels of design variables and their combinations, and measure noise and vibration response. However, it is difficult to incorporate comprehensive combinations encountered in manufacturing and assembly of the gears. For all practical purposes, the empirical work can provide trends and guidance for nominal dimensions (1). Note that such an understanding is a vital catalyst for improving transmission design and manufacturing processes. Hence, instead of a comprehensive experimental design, it may be useful to have some fundamental engineering analysis built in the experiment. For example, transmission error can be used in the experimental design rather than the individual contributors leading to it. However, it is easier to design an experiment around fundamental factors rather than derived engineering quantities like transmission error.

Transmission error is defined as the deviation of the driven gear from ideal conjugate position. Factors that contribute to the transmission error include individual dimensions of the gears like lead deviation, crown, tip and form fall-offs as well as assembly issues like misalignment. The relationship between these and transmission error can be computed (2-7). Therefore, while designing an experiment it is not necessary to include all variables that result in transmission error variation. As long as the design of experiment relates transmission error to noise, the results are applicable beyond the factors used in the experiment, but limited to those that are known to influence transmission error in general.

Besides the tolerance stack-up that results in transmission error, the effects of assembly techniques may not be that apparent. For example, a riveted versus welded carrier or a slip-fit versus press-fit shaft will exhibit considerably different noise and vibration characteristics (1). In order to study the effects of assembly techniques on noise and vibration characteristics, it is essential that the assembly techniques must be replicated to the last detail. However, the variability due to assembly techniques is difficult to control than the manufacturing variability. Hence, it is proposed that the nominal effects of the assembly techniques will be obtained empirically, and the variability effects will be determined using analytical and statistical techniques. This paper presents a method to estimate the effect of manufacturing and assembly process variability on transmission noise and vibration. In some cases, the statistical nature the manufacturing and assembly processes can be exploited to offset each other (8).

Specifically, the noise and vibration originating from the Ravigneaux type planetary gear system (forward first gear) is considered. An experiment is devised to quantify the effects of manufacturing and assembly variability on transmission noise and vibrations. Out of the five factors considered in the design of experiment, three (hole position, hole parallelism, and lead) contribute to the transmission error. The other two factors (shaft retention and bearing clearance) also contribute to the transmission error, but cannot be included in the transmission error computation due to the limitations of current transmission error computation methods. An empirical relationship between the transmission error, the bearing clearance, nature of the shaft retention (press-fit /slip-fit) and the noise/vibration is determined. The gears (sun and pinion) and carrier dimensions are measured, whereas the ring gear is assumed perfect. The transmission error estimation for planetary gear system is based on static single mesh predictions (2). The planetary system is decomposed into several single mesh systems (7) and the composite transmission error is synthesized from the single mesh predictions. The

C606/027/2002 © Ford Motor Company 2002

transmission noise and vibration empirical regression is used to illustrate the predictive capability of the model. The statistical distribution for both noise (sound pressure) and the drive shaft torsional acceleration are determined for a given tolerance and nominal values for transmission error and bearing clearance. The shaft retention is assumed press-fit.

2. TRANSMISSION ERROR COMPUTATION

Transmission error is defined as the deviation of the driven gear from its ideal conjugate position. The root-cause for transmission error is the manufacturing/assembly deviations (even though within tolerance), and the deflections due to loadings. Due to statistical nature of tooth-to-tooth or gear-to-gear variation, the transmission error varies over each cycle and each gear combination. Transmission error measurement and prediction for a single mesh two gear system has been studied extensively (2-5). Although it is difficult to predict gear noise based on transmission error, empirical regressions between the transmission error and gear noise/whine have been developed (6,9). Transmission error prediction for planetary gear systems is of recent interest (1, 8, 10). This section describes the method used to compute the Transmission error for Ravigneaux type planetary gear system. The method is based on mathematical decomposition (into several single mesh systems) and synthesis (one planetary system), similar to one used for a Simpson type planetary system by Athavale et. al. (7).

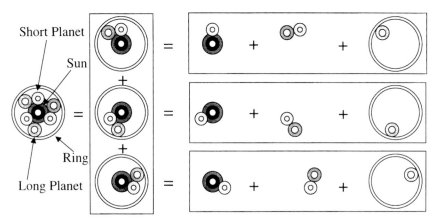

Figure 1. Decomposition and synthesis of Ravigneaux type planetary gear system.

Consider the Ravigneaux type gear system shown in Figure 1 with a sun gear, three short pinions, three long pinions, and a ring gear. In order to compute the transmission error, the planetary gear system is assumed to consist of three identical planetary systems with one short planet and one long planet (See Figure 1). The single planet system is assumed to consist of three single planet systems, namely, sun-short planet, short planet-long planet, and long planet-ring system (See Figure 1). The transmission error for these single mesh two gear systems is computed based on existing method (2). Note that the gear data used in the single mesh transmission error computation is shown in Table 1. Manufacturing and assembly errors considered in this computation are a) location of pinions, b) misalignment of pinion axes, c) lead deviation for the sun gear, d) radial bearing clearance for the pinion shaft and e) type of shaft retention. Other dimensions are at nominal values and the ring gear is assumed perfect. The transmission error results for first case (see Table 2) are shown in Figure 2. Note that the transmission error (shown over two cycles) varies within the cycle. This variation tends to accelerate and decelerate the driven gear about its nominal rotational speed, resulting in noise and vibration.

Table 1. Gear Data.

	Sun	Short Pinion	Long Pinion	Ring
Number of Teeth	31	24	25	88
Pitch Diameter (mm)	42.468	36.753	38.280	134.747
Base Diameter (mm)	44.914	34.772	36.221	125.690
Face Width (mm)	24.379	24.379	23.52	24.379
Normal Pressure Angle (degrees)	19.494	19.494	19.494	19.494
Helix Angle (degrees)	23.632 RH	23.632 LH	23.632 RH	23.632 RH
Tip Relief (mm)	0.005	0.005	0.0127	-.0.005
Form Relief (mm)	0.0025	0.0025	0.0076	0
Crown (mm)	0.0127	0.006	0.007	0.0075
Center Distance (mm)	0	43.485	47.982	0

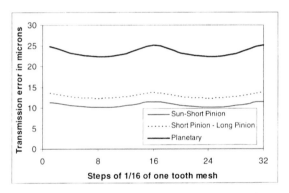

Figure 2. Transmission error computation for Ravigneaux type planetary gear system.

3. DESIGN OF EXPERIMENT

The goal of the experiment was to establish a quantitative empirical regression between the transmission error, bearing clearance, and shaft retention and the emitted noise as well as the drive shaft torsional acceleration. A two level full factorial design will include two levels of each of these factors. However, it is difficult to obtain precisely two levels of transmission error because several factors contribute to transmission error. The source of transmission error was limited to the radial location of pinion axes holes, misalignment of the pinion shaft and the lead deviation for the sun. A fractional factorial experimental design using shown in Table 2 is used for the noise and vibration measurement. Note that even this design will lead to multiple levels of the transmission error. The gears and the carrier were measured to

C606/027/2002 © Ford Motor Company 2002

confirm not only that the relevant dimensions conform to the experimental design but also the rest are at or near the nominal dimensions (See Table 1).

The experimental set-up shown in Figure 3 is a standard set-up used for such measurements (1, 11). The transmission is enclosed in an acoustic room to prevent echoes, and sound interference. The sound pressure level is measured using a microphone that is directly placed above the planetary gear system. It is held at 228.6 mm above the transmission casing. The drive shaft acceleration is measured using two tangential accelerometers placed in the coupling between drive shaft and the shaft of the load motor. The input shaft is connected to an electric motor that drives the transmission. The noise and vibration levels of interest are for the gear configuration in forward first gear. Before the measurements are taken, the transmission undergoes complete warm-up. The warm-up includes running the transmission at 700 rpm until the oil bath temperature reaches 65.6° C. After completion of the warm-up, the measurements are recorded continuously as the rpm is increased to 4000. The measurements are repeated five times. Figures 4 and 5 show the average measurements for the six cases (Table 2). The resonance peaks of interest occur at 820 Hz and 690 Hz, respectively. The sound pressure levels at these peaks will be used to develop the sound and vibration empirical regression, described in the next section.

Table 2. Design of Experiment.

Case Number	Hole Position	Misalignment	Lead	Bearing Clearance	Shaft Retention
1	+	-	+	+	+
2	+	-	+	-	+
3	+	+	-	+	-
4	-	-	-	+	-
5	-	+	+	-	-
6	-	+	-	-	+
	+ 0.1016 mm	+ 0.0254 mm	+ 0.0457 mm	+ 0.0420 mm	+ Press fit
	- 0.0127 mm	- 0.0051 mm	- 0.0051 mm	- 0.0180 mm	- Slip Fit

Figure 3. Experimental set-up.

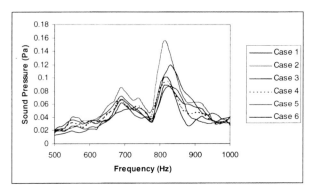

Figure 4. Sound pressure measurement.

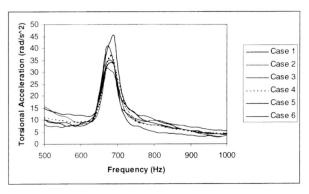

Figure 5. Accelerometer signature.

4. RESULTS AND ANALYSIS

The goal of this work is to establish empirical regression equations between the manufacturing and assembly variations, and the transmission noise and vibration. In this particular experiment, three independent factors are used for developing the empirical regression, namely, the peak-to-peak transmission error, bearing clearance for pinions and the pinion shaft retention. Note that the transmission error is a derived quantity and is determined from the measured dimensions as explained earlier. As such, the transmission error will depend on the variation of several dimensions (2, 3, 5, 7, 10, 11). However, the scope of variables that influence the transmission error is limited to the radial location of pinion shaft, misalignment of pinion shaft and lead deviations for sun. Note that the lead deviations and misalignment may add or cancel depending on the directions. Once the transmission error is computed, a least squares fit can be used to determine the empirical equations. Note that if the transmission error was constant over the meshing cycle then the gear noise can be solely attributed to gear impact at approach and recess. In this case, the variation in transmission error over a cycle and associated noise is of interest. Therefore, peak-to-peak transmission (PTE) error will be used in the regression. In addition, the shaft retention (SR) can be either slip-fit or press-fit (1 or –1). Hence, during predictions, either of the states will be used. The

C606/027/2002 © Ford Motor Company 2002

bearing clearance (BC) is obtained from the dimensional stack-up and used in the least squares fit. The empirical regressions obtained in this fashion are listed in Equations 1 and 2.

$$\text{Sound Pressure (Pa)} = 0.000639*PTE - 1.8109*BC + 1.6101*SR - 3.082 \qquad (1)$$
$$\text{Acceleration (rad/s2)} = -6*PTE - 103*BC + 1368*SR - 3800 \qquad (2)$$

The predictions based on these empirical regressions and the three sigma limits are shown in Figure 6. The deviation in measurement is obtained from the standard deviation of the five replications conducted for each test. Further, the prediction error is obtained from the standard least squares analysis. As these figures indicate, the empirical regression shown above is quite reasonable. Note that case 2 is worst for sound pressure, but it is the best for drive shaft acceleration. On the other hand, cases 1 and 6 are better for either the sound pressure or the acceleration. This observation indicates that the primary factor that contributes to the sound and vibration under these circumstances is the shaft retention i.e. the press-fit is better. However, case 2, which has the press-fit shafts, exhibits minimum acceleration and maximum sound pressure. Thus, the relationship between these variables is not that intuitive.

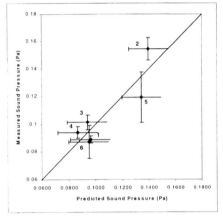 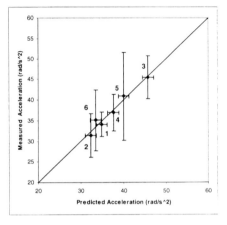

(a) Sound pressure   (b) Drive shaft acceleration

Figure 6. Predictions from empirical regressions vs. measurements.

Once the empirical regression is established between the transmission error, bearing clearance, and shaft retention, and the noise and vibration it can be used for a sensitivity analysis. Note that even though the transmission error source was constrained to few factors, the empirical regression should be independent of this source as long as the final transmission error outcome is the same. Let us assume input specification for a press-fit shaft in terms of the transmission error and bearing clearance to be 2.8 microns and 0.03 mm. The bivariate distribution about these normal values is shown in Figure 7. This represents a classic symmetric distribution. The predictions obtained for the sound and vibration are shown in Figure 8. Although, overall distribution looks normal, it is skewed with respect to the horizontal axes. This indicates an empirical regression between the sound and vibration response. The feasible region for a robust design should be elliptical (12), rather than rectangular as shown in the Figure 8. Also, note that the bottom plane is also skewed in the vertical directions indicating lower probability of occurrence at the lower values of sound and vibration.

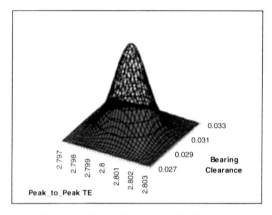

Figure 7. Input bivariate distribution.

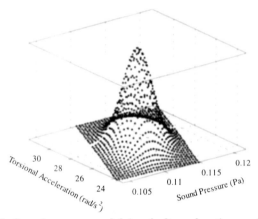

Figure 8. Sound pressure and drive shaft acceleration predictions.

5. SUMMARY

The empirical regression equations between transmission error, shaft assembly, bearing clearance, and transmission noise and vibration were developed for the Ravigneaux type planetary gear system. The transmission error computation for the Ravigneaux type planetary gear system was also developed. The sources for transmission error were constrained to lead deviation, pinion hole position and parallelism. The empirical regression was used to illustrate the sensitivity of the output response to a predefined bivariate normal distribution of transmission error and bearing clearance.

C606/027/2002 © Ford Motor Company 2002

6. ACKNOWLEDGEMENTS

The authors wish to acknowledge S. Hassan, W. VanHafften and the Hemi-cell test team for their assistance in conducting the tests.

7. REFERRENCES

(1) Athavale, S. M., Gardner, G. D. and Trent, M., *"Influence of Assembly Techniques, Misalignment Errors, and Manufacturing Variation on Noise and Vibration Characteristics of Automatic Transmission,"* Submitted for review to Global Powertrain Engineering Conference, 2002.

(2) Smith, J. D., *"Gear Noise and Vibration,"* ISBN: 0-8247-6005-0, 1999.

(3) Dunn, A. L., Houser, D. R., and Lim, T. C., *"Methods for Researching Gear Whine in Automotive Transaxles,"* Proceedings of the 1999 Noise and Vibration Conference, SAE Technical Paper 1999-01-1768, 1999.

(4) Mark, W. D., *"Analysis of the Vibratory Excitation of Gear Systems: Basic Theory,"* Journal of Acoustical Society of America, 63(5), pp. 1409-1430, 1978.

(5) Mark, W. D., *"Analysis of the Vibratory Excitation of Gear Systems II: Tooth Error Representations, Approximations, and Application,"* Journal of Acoustical Society of America, 63(5), pp. 1409-1430, 1978.

(6) Chung, C-H., Steyer, G., Abe, T., Clapper, M., and Shah, C. *"Gear Noise Reduction through Transmission Error Control and Gear Blank Dynamic Tuning,"* Proceedings of the 1999 Noise and Vibration Conference, SAE Technical Paper 1999-01-1766, 1999.

(7) Athavale, S. M., Krishnaswami, R., and Kuo, E. Y., *"Estimation of Statistical Distribution of Composite Manufactured Transmission Error, A Precursor to Gear Whine, for A Helical Planetary Gear System,"* Proceedings of 2001 SAE Noise & Vibration Conference & Exposition, April 30-May3, 2001.

(8) Athavale, S. M., *"Assembly Strategies to Reduce Gear Whine in Planetary Transmissions,"* Technical Paper MS02-186, SME, Transactions of NAMRI, 2002.

(9) Lorea, A., Mora, G., and Ruspa, G., *"Advanced Statistical Methods For The Correlation Between Noise And Transmission Error In Gears,"* Proceedings of 21st FISITA Congress, paper number 865144, pp. 2.285-2.290, 1986.

(10) Ishida, T. and Hidaka, T., "Effects of Assembly And Manufacturing Errors On Transmission Error Of Planetary Gears," Proceedings of International Power Transmission and Gearing Conference, ASME DE-Vol. 43-1, pp. 375-381, 1992.

(11) Oswald, F. B., Townsend, D. P., Valco, M. J., Spencer, R. H., Drago, R. J., and Lenski Jr., J. W., *"Influence Of Gear Design On Gearbox Radiated Noise,"* Gear Technology, Vol. 15(1), pp. 10-15, 1998.

(12) Yu, J-C., and Ishii, K., *"Design For Robustness Based On Manufacturing Variation Patterns,"* Journal of Mechanical Design, ASME, Vol. 120, pp. 196-202, 1998.

C606/004/2002

The improvement of vehicle performance using predictive diagnostic algorithms

L BORGARELLO, M GAMBERRA, and **L GORTAN**
Statistical Methods, Fiat Research Centre, Torino, Italy

ABSTRACT

The goal of predictive diagnosis is **to identify and make explicit a slow degradation (or wear-out) as much as possible in advance of the failure itself**. In this article a simple statistical approach to predictive diagnosis is described on an application regards the automotive sector and in particular the application of predictive diagnostic algorithms to the monitoring of the traction battery performances. The application was developed within the Atena project (Ambiente Traffico TElematica Napoli) where an experimental eco-telematic fleet of 80 vehicles was used.

1 INTRODUCTION

A lot of work has already been done in the development of fleet monitoring systems for heavy-duty vehicles and particularly for public transportation. These systems are able to collect many data about the service and operating conditions of the vehicle, such as fuel tank level and other service-dependent parameters (kilometres, average speed, driver identification and so forth). In some cases they are also able to call a remote service station, in case of failure, and support remote diagnosis. This is achieved by transferring diagnostic codes from vehicle OBD (On-board Diagnostic) system to the remote service station.

In particular, some public transport operators are already installing a sophisticated wire-less monitoring system providing extensive operating and diagnostic data collection capabilities as a standard feature of each new bus being set in operation since this year. This monitoring system include the ability to collect the values of all "normal production" sensors installed on the vehicle. Additional sensors are provided for detecting early degradation in two maintenance critical vehicle auxiliaries (engine cooling system and compressed air generation).

2 PREDICTIVE DIAGNOSIS

Nowadays maintenance actions are based on periodical controls and on management of fault events. Very limited experience exists on the use of data collected by fleet monitoring system in view of implementing "on-condition" preventive maintenance strategies.

On-board diagnosis is a part of the software control present in most modern automotive Electronic Control Units. Current OBD approaches are based on the development of qualitative and deterministic models. As defined recently, "the goal of OBD algorithms is to detect faulty conditions in the system, identify the incorrect or missing functions and actuate the most appropriate recovery action in order to guarantee passenger safety, avoid engine damages, reduce emission and so on". Therefore the goal of these algorithms is **to identify and make explicit a failure once this has occurred rather than detect degradation (or wear-out) as much as possible in advance of the failure itself**.

So far the development of a statistical approach to predictive diagnosis has been hindered by the difficulty to tune the algorithms, as this require a large number of sample data which are very expensive and time demanding until a wire-less fleet monitoring system is installed.

On the other hand modern vehicles are equipped with a larger number of sensors capable to monitor many vehicle conditions and environmental parameters. Moreover the field of the telemetry has steadily increased recently and the transmission of large data is no longer seen as a serious bottleneck. These two conditions, which contemporarily exist, make possible and opportune the exploitation of this kind of information, which until now has been completely lost.

There are however some difficulties arising:
- Each failure mode of the single component has its own cause and it must be modelled with a specific approach. This brings to an increase of the overall complexity.
- The degradation of the performance over time could have a extremely slow trend or a quite unpredictable behaviour. This makes the model estimation a challenging task.
- Each predictive model can depend, in principle, on a large number of parameters. These explicative parameters can be directly controllable (for example the injection time) or non controllable (for example cooling water temperature) and finally depending on the external environment (for example outside air temperature, mission profile, etc.). Each one of these parameters can, in principle, have a determinant contribution in explaining the behaviour of the component over time. A multivariate modelling approach is therefore strictly necessary.
- Last but not least, the database available from a continuously monitored fleet of public transport tends quickly to become unmanageably huge. On the other hand, a large experimental base is necessary to increase model predictivity and to validate it in working conditions which are to be as different as possible.

A statistical approach can overcome the above mentioned problems: specific statistical techniques exists in order to:
- adequately model over time a signal (i.e. system performance and time series analysis);
- adequately deal with multivariate models (general linear models);
- reduce, if necessary, problem dimension and complexity, by keeping at the same time, as much information as possible (multivariate data analysis like principal component analysis and so on).

Advanced and well settled statistical modelling techniques exist in order to provide the necessary support in the analysis of this large amount of data. Specific topics deal indeed with the field of time series analysis, statistical process control (SPC), multivariate modelling (like general linear models) and multivariate analysis (like, for example, principal component analysis).

Recent applications of statistical process control techniques for the definition of predictive diagnostic algorithms already gave good results. In the next paragraph we will describe in further details the application of a predictive algorithms on the degradation of a traction battery in use by the Naples municipality.

3 APPLICATION DOMAIN AND ADVANTAGES

The development and on line use of Diagnostic Algorithms find its natural domain in the management of big public fleet, like public transport. This domain is potentially huge: the total public road transportation fleet in Western Europe (total buses over 3,5 t) consists of about 410.000 vehicles, and the average mileage for urban bus service is 50.000 km. Vehicle maintenance, moreover, has a significant cost (in the range of 0.5 Euro/km, or 25,000 Euro/vehicle/year for conventional 12 meter urban buses).

Regarding the vehicles subsystem / component that could benefit from predictive diagnosis, it is clear that these typology of component are most suited:
- components related to environmental performances, like exhausted gas treatment systems
- component with a short and non clearly predictable life (like battery and traction battery in electric vehicles, for example)
- critical component in term of safety for the passengers
- critical components in term of maintenance costs

In addition, a limited set of overall vehicle operation data is also collected. Regarding fuel conversion efficiency, commercially available systems allow accurate measurement of overall fuel consumption on a mileage basis. In some advanced systems, fuel consumption analysis with respect to vehicle/engine operating modes is also provided (e.g. fuel consumption at idle) the innovation of the project shall consist in the introduction of a algorithm able to analyse actual performance with a reference energy model and detect deviations.

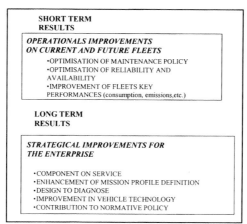

Fig. 1 Long/short term expected advantages of predictive diagnosis

4 AN APPLICATION

4.1 Introduction

The Atena project (Ambiente Traffico TElematica Napoli) presents an ideal scenario for this type of analysis due to the use of an experimental eco-telematic fleet of 80 vehicles at low environmental impact used from normal citizens and personnel of Naples municipality for a period of three years.

Every vehicle is equipped with an on-board telematic system that allows the data acquisition of engine, vehicle and geographic variables, the memorisation and the transmission to a control station using a store-and forward methodology. The data acquisition system was unattended. The data acquisition is switched on automatically at the engine start. The acquisition ends at the following key-off. Every acquisition covers a specific mission of the vehicle and it is possible, due the acquisition rate up to 2 Hz, the data store up to 8 hours of missions. The data stored in the on-board system were periodically forwarded to the control station when the on-board memory was occupied more than 50%. The native format was compressed, crypted and sent to the ground station via the GSM cellular network. A specific software will convert data in order to be compatible with commercial software like SAS^{TM} or Matlab^{TM} applications.

In particular, for the predictive algorithms has been interesting the study of 25 Electric cars (Fiat Seicento) equipped with two different types of batteries: OVONIC and EXIDE.

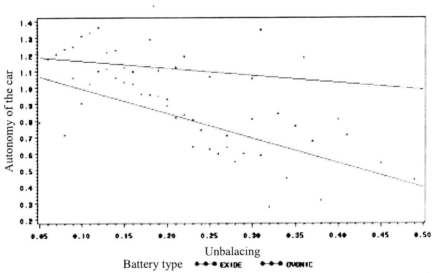

Figure 2 Link between disequalisation level of the battery and performance (autonomy)

For these cars there has been observed a correlation between the unbalancing of batteries and their performances. As it clearly emerges in Figure 2, the negative trends (red for exide and green for ovonic) emphasise the link between the autonomy of the car (mean value in term of km/Ah) and the disequalisation level.

From this graph it is evident that it is important to prevent, especially for EXIDE batteries, an high unbalacing. This is the aim of the predictive diagnostic algorithm, that will be described in the next paragraph.

4.2 Description of the application
In order to prevent situation of unbalancing it has been observed the relationship between the variation of this variable over time and problems to the traction system from the user point of view. The red lines represent indeed the battery substitutions and the green line represent the call to the green number (when the car doesn't work properly).

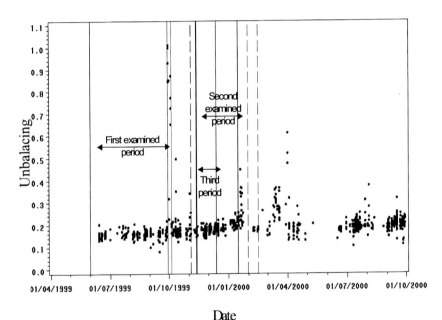

Figure 3 Behaviour of unbalancing over time

It possible to observe that, before a green or a red line, sometimes it is possible to observe a gradual increment in unbalancing. In order to find as soon as possible an increasing trend of this parameter have been used some Statistical Process Control methods; and good results have been found with the application of a Cusum Chart modified in order to solve the following problems:
- lack of a nominal value for the examined parameter
- sampling non regular in time
- number of measures non equal in different sampling.

For example in Figure 4 the application of the method in three different period of the considered time-history are reported.

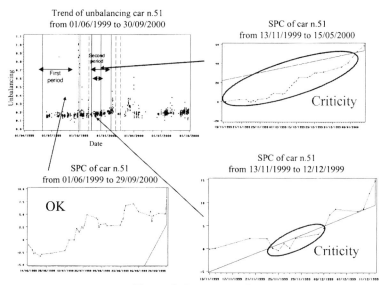

Figure 4: Some examples

It is important to observe that:

- in the first period no trend has been observed but probably the following increment could due to the battery substitution
- the second increment in clearly observable not only some day before the rapid increase (period 2) but also one month before (period 3). In Figure 5 is reported another example of good forecast.

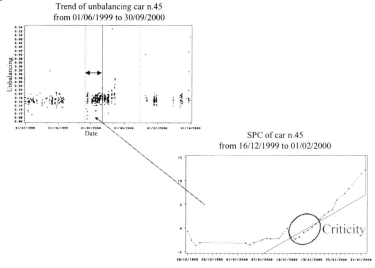

Figure 5 An example of slow degradation correctly identified by the algorithm

                                               C606/004/2002 © IMechE 2002

The methodology could be implemented (during the phase of charging or discharging) to take under control the batteries and to organise a maintenance of balancing in order to prevent an high unbalancing that, not only decrease the performances of the batteries, but that is also dangerous for batteries itself.

5 CONCLUSIONS

The goal of predictive diagnosis **to identify and make explicit a slow degradation (or wear-out) as much as possible in advance of the failure itself**. In this article a simple statistical approach to predictive diagnosis was presented. The application regards the automotive sector and in particular the application of predictive diagnostic algorithms to the monitoring of the traction battery performances. The scenario was identified by the Atena project (Ambiente Traffico TElematica Napoli) due to the use of an experimental eco-telematic fleet of 80 vehicles.

This application proves the potential advantages that could be reaped with this approach. It is possible, indeed, to improve the system reliability, safety and availability; the improved knowledge of the vehicles mission profiles, moreover, will greatly increase the efficiency of the design of new products, leading to a large increase in industry competitivity.

Requirements for the successful applicability of this approach are the availability of a wireless fleet monitoring system techniques and data mining and statistical analysis capability.

6 REFERENCES

M. André, A.J. Hickman, D.Hassel, R. Joumard "Driving Cycles for Emission Measurements Under European condition" Rif 950926 SAE

M. André, "Driving Cycles Development: Characterization of the Methods" Rif 961112 SAE

G.L. Berta, P.Casoli "Development of compact driving cycles representative of urban mission" Florence Ata 1999: Ref. 99A4053

L. Borgarello, R. Fontana, A. Fortunato, L. Mina: "Identication of driving cycles and emission in the traffic of Bologna" Rif. 01A1012 Florence ATA 2001

L. Borgarello, R. Fontana, A. Fortunato, L. Mina: "Determinazione sperimentale delle emissioni provenienti da autoveicoli circolanti in ambiente urbano", Bologna, 1999

L. Della Ragione, A. Buonocore, M. Rapone "Metodi multivariati per l'analisi e visualizzazione di dati sperimentali rappresentativi dell'utilizzo reale di un'autovettura sulla tangenziale di Napoli" SUGItalia '98: Atti del Convegno

W.R. Dillon, M.Goldstein "Multivariate Analysis" John Wiley & Sons

G. Ferulano, L. Gortan, A. Sforza, D. Tartaro :"The ATENA Project - Ambiente, Traffico, Telematica Napoli " Paper 2399 - Intelligent Transport System - World Congress - Turin 6 - 9 nov 2000.

J.D. Jobson, "Applied Multivariate Data Analsisys" Springer-Verlag New York

M.Rapone, L.Della Ragione, F. D'Aniello, V. Luzar: "Experimental Evaluation of Fuel Consumption and Emissions in Congested Urban Traffic" Rif 952401 SAE

C606/018/2002

Reliability modelling using warranty data

I F CAMPEAN
School of Engineering, University of Bradford, UK
D BRUNSON
Powertrain Quality and Reliability, Jaguar Cars, Coventry, UK

ABSTRACT

This paper discusses hazard rate modelling and reliability prediction based on automobile warranty failure data. Two case studies are used to illustrate the practical application of the method. The first case study looks at modelling early field failure data (4-5 months in service) to identify components that have the potential of becoming actionable items (such as a recall decision). The second case study investigates the reliability of a component for which both time to failure and mileage to failure are of interest.

INTRODUCTION

Collection and analysis of vehicle field failure data is very important as it reflects the true reliability performance of automotive components, systems and of the vehicle as a whole, when operated under various duties, stresses and environmental conditions which define the customer usage. Automotive warranty covers a pre-determined period of time and/or mileage accumulation (e.g. 12 months / unlimited mileage, or 36 months / 60,000 miles, whichever comes first) during which costs incurred by customer claims are met by the manufacturer. Automotive manufacturers maintain warranty databases in which details of warranty claims are recorded.

However, reliability modelling based on warranty data is complicated by censoring. From a statistical point of view, field failure data samples extracted from automotive warranty databases can be considered as randomly censored (1), typically with unknown censoring times. A generic problem is that warranty information is restricted to failure events occurring inside the warranty period and very little or no information (such as mileage accumulated to date) is readily available for vehicles that have not experienced any concern (no warranty claim) with the component or system under analysis. The situation complicates further if the warranty plan has both time and mileage limits, which is a typical situation for the US market. For example, if the warranty plan covers 36 months in service or 36000 miles, whichever occurs first, failures of cars exceeding 36000 miles and times in service less than 36 months are not recorded.

One common approach addressing difficulties with the statistical analysis of this type of data consists of using additional information and assumptions about the censoring process, which in the case of automotive warranty data consists of the mileage accumulation process. Starting with the pioneering work of Suzuki (2, 3), there has been a great deal of development in this direction in recent years (4 - 7).

The way in which warranty data modelling is performed depends on the objectives of such analysis. Typical reasons for carrying out warranty data analysis are the following:

- Forecasting warranty claims (8, 9);
- Risk assessment and monitoring, for early identification of systems and components that show a potential of becoming actionable items (i.e. to determine a vehicle recall) (10).
- Reliability assessment / modelling for components and subsystems.

If the aim of the analysis is to forecast warranty claims, then the assessment of the joint distribution of failure and mileage accumulation is of interest (8). Other approaches (9) rely on time series analysis or nonparametric modelling using artificial neural networks.

From an engineering point of view, hazard analysis is a preferred practical tool for analysing reliability data. The technique is generally well understood by engineers, it requires a reasonable amount of calculation and the common graphical representations – hazard rate and cumulated hazard plots, can be used to extrapolate trends and make conclusions about the reliability of the component, useful in engineering terms. However, the approach cannot be easily applied to randomly censored data and when information is missing, as it is generally the case with warranty failure data. Other analytical methods often used in practice – like Weibull analysis, have similar difficulties in coping with this type of failure data.

If the censoring distribution is known, this can be used together with failure data to perform hazard analysis. The basic approach is discussed by Lu (10), who used the distribution of the mileage accumulation process to estimate the expected number of censoring for each mileage interval. A more refined approach was presented in (11), where a MLE (Maximum Likelihood Estimation) type estimate of the interval hazard rate is considered in conjunction with a Bayesian smoothing technique to perform hazard rate analysis. This latter approach will be discussed in the following and its practical application illustrated with two case studies.

METHODOLOGY

Warranty failure data
Warranty data is typically presented in the format presented in Table 1 (MIS – *Months In Service*). This is based on the fact that the mileage accumulated by the vehicle is recorded with any warranty claim. Failure information ($R_{i,j}$) is given either as number of failures or as repairs per 1000 vehicles, which in turn can be then converted, for convenience, into number of failures.

Table 1. Structure of failure warranty data

MIS	Sales / field volume	Mileage [up to k miles]			
		1	2	3	...
1	N_1	$R_{1,1}$	$R_{1,2}$	$R_{1,3}$	...
2	N_2	$R_{2,1}$	$R_{2,2}$		...
$\vdots$	$\vdots$	$\vdots$			
M	N_M	$R_{M,1}$		...	

Censorings are vehicles that have only been in the field for m ($m < M$) MIS or have been in the field for M MIS but have not experienced failures with the component under investigation. The mileage accumulated by the censored vehicles is typically unknown.

A general assumption made with respect to the censoring process is that censorings have identical statistical properties with the lifetimes, i.e. the censored vehicles belong to the same statistical population as the ones that have experienced the component failure. A further simplifying assumption commonly made is that the censoring process is non-informative (7), i.e. the censoring and failure distributions are statistically independent (which is reasonable where the time to repair is unlikely to influence the mileage accumulation process).

Mileage accumulation process
It is generally accepted that the mileage accumulation process can be modelled by a lognormal distribution (4, 10, 11, 12), Equation 1. Parameters of the lognormal distribution can be estimated using the sample mean ($\overline{X}$) and standard deviation (s), from customer mileage accumulation data samples (see Equations 2 and 3).

$$f(t;\mu,\sigma)=\frac{1}{\sigma \cdot t \cdot \sqrt{2 \cdot \pi}} \cdot \exp\left[-\frac{1}{2}\cdot\left(\frac{\ln t-\mu}{\sigma}\right)^2\right] \tag{1}$$

$$\hat{\mu}=\ln(\overline{X})-\frac{1}{2}\cdot\ln\left[1+\left(\frac{s}{\overline{X}}\right)^2\right] \tag{2}$$

$$\hat{\sigma}^2=\ln\left[1+\left(\frac{s}{\overline{X}}\right)^2\right] \tag{3}$$

The mileage distribution is vehicle and market specific, and should be estimated from additional customer surveys or follow-up studies (1, 6, 11). The mileage information associated with warranty failure data is not suitable for the assessment of mileage distribution, as failure might be favoured by a specific usage pattern, such as low or high mileage accumulation.

Lu (10) has reported a comprehensive study on mileage accumulation based on customer surveys for passenger car and light duty trucks for the US market. This analysis has shown that the average monthly mileage accumulation for a passenger car is around 1000 miles. Another interesting finding was that the ratio between the standard deviation and the mean was around 0.5 (0.4 – 0.6) across the samples examined. This conclusion was subsequently validated by other studies based of warranty failure data from different markets, see for example (11).

Hazard rate function

A useful approximation of the hazard rate function is obtained by a piecewise constant hazard rate assumption (equation 4).

$$h(t) = \{h_j, t_{j-1} \leq t < t_j \tag{4}$$

where h_j denotes the j-th interval hazard rate, assumed constant. Taking censoring into account, a maximum likelihood type estimate for a mileage interval hazard rate can be derived as follows:

$$\hat{h}_j = \left[\frac{\text{Number of failures in current interval}}{\text{Mileage accumulated by all vehicles in current interval}} \right]_j = \frac{f_j}{M_{\Sigma j}} \tag{5}$$

where $M_{\Sigma j}$ is the sum of:

i)   mileage accumulated (from the beginning of the mileage interval) by the f_j vehicles which have failed in the respective mileage interval, until the occurrence of their failure;

ii)   mileage accumulated (from the beginning of the mileage interval) by the c_j vehicles censored within the interval, until censoring;

iii) mileage accumulated by the vehicles surviving the mileage interval; for each unit this is equal to the length of the mileage interval.

The problem is that the number of censoring for each mileage interval is not available. If the mileage distribution is known, the expected number of censoring for the j-th mileage interval from the n_i vehicles censored at i MIS can be calculated as follows:

$$c_{i,j} = n_i \cdot \left[\Phi(j \cdot u; i \cdot \mu, \sigma) - \Phi((j-1) \cdot u; i \cdot \mu, \sigma) \right], \tag{6}$$

where $\Phi(t; \mu, \sigma)$ is the lognormal cdf and u is the length of the mileage interval. If the exact mileage at failure or censoring is unknown or not available, the usual approach is to consider that failure or censoring occurred at the middle of the interval.

Bayesian smoothed hazard rate function

A common problem with plotting interval hazard rates which are obtained from heavily censored failure data (such as field failure warranty data), consists of the fact that it usually results in an uneven disjointed contour which is difficult to interpret by the engineers. A way around this problem is to use a Bayesian smoothing technique (13), which can be applied to obtain a more robust estimation for the interval hazard rate.

The underlying idea of this Bayesian smoothing technique is that significant variation in the hazard rate between adjacent mileage intervals is unlikely, therefore a functional relationship between consecutive interval hazard rates can be assumed. With the further assumption of a conjugated gamma prior pdf, $g(a, b)$, the Bayesian estimate for the j-th mileage interval hazard rate can be written as follows:

$$\hat{h}_j = \frac{\alpha_j^*}{\beta_j^*} = \frac{\alpha_j + f_j}{\beta_j + M_{\Sigma j}} , \tag{7}$$

where $g(a_j, b_j)$ is the Gamma prior *pdf* for the *j*-th interval hazard rate and $g^*(\alpha_j^*, \beta_j^*)$ is the posterior hazard rate distribution, with $\alpha_j^* = \alpha_j + d_j$ and $\beta_j^* = \beta_j + M_{\Sigma j}$. The prior distribution for h_j is related to the posterior distribution of h_{j-1} as follows:

$$\alpha_j = k_j \cdot \alpha_{j-1}^*; \ \beta_j = k_j \cdot \beta_{j-1}^* \tag{8}$$

where $k_j \leq 1$ is a coefficient that controls the passage of information through adjacent intervals. Evaluation of the prior parameters should be coherent with the formal requirements for the stochastic random walk model, i.e. mean preservation and increased uncertainty with time (13). By imposing these requirements and based on some asymptotic approximations (13), the coefficients k_j can be derived as follows:

$$k_j = \left(\alpha_{j-1}^*\right)^2 \Big/ \left[\left(\alpha_{j-1}^*\right)^2 + D_f^{-1} - 1\right] \tag{9}$$

where D_f is a discount factor subjectively chosen to denote the fraction of information that passes through the time intervals. It is easily verified that if $D_f \rightarrow 1$ then $k_j \rightarrow 1$, hence there is no loss of information, in the sense that in any interval the posterior distribution is the prior distribution for the next interval hazard rate. Whilst if $D_f \rightarrow 0$ then $k_j \rightarrow 0$ and no information passes through the time intervals, hence the Bayesian estimator is reduced to the *MLE*. An iterative process can be easily established to perform these calculations (11), as follows:

1) The *MLE* hazard rate estimate for the first mileage interval is used to initiate the Bayesian algorithm (i.e. non-informative prior distribution for the first interval hazard rate),

$$\hat{h}_1 = f_1/M_{\Sigma 1} ,$$

2) For all subsequent intervals $j = 2..n$ (*n* – number of intervals):
   - Calculate k_j using Equation 9.
   - Calculate α_j and β_j using Equation 8.
   - Calculate α_j^* and β_j^* and the *j*-th interval hazard rate estimate ($\hat{h}$) using Equation 7.

The way in which the Bayesian smoothing works essentially depends on the value chosen for the discount factor D_f. A deeper discussion on this issue based on a sensitivity analysis is provided in (14). However, for practical reasons a conservative (i.e. low, $D_f = 0.1 - 0.3$) value can be chosen.

Reliability predictions
Based on the constant piecewise hazard rate function, a piecewise exponential estimator for the survival function can be defined, given by Equation 10. Kim & Proschan (15) have argued that the piecewise exponential estimator (*PEXE*) of the survival function (Equation 7) is, under mild regularity conditions, asymptotically equivalent to the Kaplan-Meier estimator (16) of the survival function. Furthermore, the *PEXE* has other attractive features, like continuity, age sensitiveness with respect to the times to censoring, flexibility with respect to grouped data and interval endpoints. In practice, survival probability predictions can be easily calculated for the

mileage interval endpoints from Equation (10) and based on interval hazard rate estimates, using the recurrent Equation 11.

$$
S(t) = \begin{cases} e^{-h_1 \cdot t}, \text{ for } 0 \le t < m_1 \\ e^{-[h_1 \cdot m_1 + h_2 \cdot (t - m_1)]}, \text{ for } m_1 \le t < m_2 \\ e^{-\left[h_1 \cdot m_1 + h_j \cdot (t - m_{j-1}) + \sum_{i=1}^{j-1} h_i \cdot (m_i - m_{i-1}) \right]}, \text{ for } m_{j-1} \le t < m_j, j = 1..n \end{cases}
\tag{10}
$$

$$
\hat{S}_j = \begin{cases} e^{-\hat{h}_1 \cdot m_1}, \text{ for } j = 1 \\ \hat{S}_{j-1} \cdot e^{-\hat{h}_j \cdot (m_j - m_{j-1})}, \text{ for } j = 2..n \end{cases}
\tag{11}
$$

EXAMPLE 1: HAZARD ANALYSIS OF EARLY FIELD FAILURE WARRANTY DATA

A case study of early field failure warranty data analysis was presented by Lu (10), which focused on identifying issues that have the potential of becoming actionable items (such as a recall decision). The data consisted of warranty failure records for a component "C" from production months between July to November 1994 (November production month was considered 1 MIS, October – 2 MIS, …, and July – 6 MIS). Lu (10) estimated the average monthly mileage accumulation to 1000 miles, with the standard deviation equal to half the expected mean. Using a lognormal distribution for the mileage accumulation process, the distribution of the censored vehicles with mileage intervals for each production month was calculated using Equation (6). The data is summarised in Table 2, where the last 2 columns give the cumulative number of failures and censorings for each interval for production months between July and November.

From an engineering point of view, a concern is raised with the reliability of an automotive component if early field failure data indicates either a prematurely and severely increasing hazard rate pattern or a hazard rate that is significantly higher than what was expected. Both scenarios would result in failure probabilities significantly higher than expected (or acceptable), meaning that there is a high potential risk that many customers would experience failures.

The reliability prediction method employed by Lu [8] consisted of using a non-parametric cumulated hazard estimator to perform Weibull regression analysis. However, the results (Weibull plots and estimates) are somewhat inconclusive, in particular with respect to November production month, which shows different hazard rate pattern (increasing hazard rate) and reliability predictions (characteristic life about 50 times lower than that for production months between July and October).

The case study presented by Lu (10) was reanalysed using the method described in this paper. Figure 1 shows two hazard rate plots for the August (4 MIS) production month failure data (the August production month was chosen to illustrate the hazard rate plots as it had the largest population of vehicles in the field and provided the most consistent failure data sample). The stepped contour in Figure 1 represents the hazard plot obtained by using the maximum likelihood estimator for the interval hazard rate given by equation (6), whereas the dotted contour is the hazard plot obtained from using the Bayesian smoothed hazard rate estimator (a continuous representation was preferred for the Bayesian smoothed hazard rate plot to emphasise the hazard trend). The value of the discount factor D_f was set to 0.25.

Table 2. Warranty failure data (10)

Mileage band	July (5 MIS)		August (4 MIS)		September (3 MIS)		October (2 MIS)		November (1 MIS)		July – Nov (all data)	
	Fail	Cens	Fail	Cens	Fail	Cens	Fail	Cens	Fail	Cens	Fail	Cens
0 – 1	2	5	21	123	36	598	29	3631	7	17658	95	22015
1 – 2	1	277	4	3715	30	8152	18	16130	10	10705	63	38979
2 – 3	1	967	8	8653	13	10700	10	8881	5	1170	37	30371
3 – 4	0	1310	4	8396	8	6785	9	3100	1	131	22	19722
4 – 5	0	1197	2	5840	10	3411	3	997	0	21	15	11466
5 – 6	1	869	2	3547	4	1596	0	312	0	3	7	6327
6 – 7	1	609	3	2110	2	765	0	110	0	0	6	3594
7 – 8	0	390	1	1166	0	353	0	37	0	0	1	1946
8 – 9	1	252	1	667	1	170	0	17	0	0	3	1106
9 – 10	0	157	0	386	0	82	0	7	0	0	0	632
10 – 11	0	104	1	207	0	43	0	1	0	0	1	355
11 – 12	0	62	1	123	0	20	0	0	0	0	1	205
+ 12	0	114	0	183	0	26	0	0	0	0	0	323
Total	7	6313	48	35116	104	32701	69	33223	23	29688	251	137041

From Figure 1 it can be seen that the hazard rate pattern (as indicated by the Bayesian smoothed hazard plot) decreases slightly up to 1000 miles in service and it is constant thereafter ($\cong 3.3 \ 10^{-4}$, see the table attached to Figure 3).

Mileage (k miles)	August		$M_{\Sigma j}$	Hazard rate	
	Fail	Cens		MLE	Bayesian
0 – 1	21	123	35135.3	0.00060	0.00060
1 – 2	4	3715	33703.5	0.00012	0.00036
2 – 3	8	8653	27176.1	0.00029	0.00034
3 – 4	4	8396	18309.0	0.00022	0.00032
4 – 5	2	5840	11131.6	0.00018	0.00031
5 – 6	2	3547	6452.5	0.00031	0.00031
6 – 7	3	2110	3672.9	0.00082	0.00032
7 – 8	1	1166	2086.3	0.00048	0.00033
8 – 9	1	667	1192.1	0.00084	0.00033
9 – 10	0	386	688.7	0.00000	0.00033
10 – 11	1	207	402.8	0.00248	0.00034
11 – 12	1	123	238.0	0.00420	0.00034
+ 12	0	183			

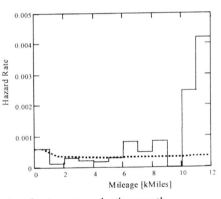

Figure 1 Hazard rate calculation and plots for August production month

In Figure 2 are shown comparatively the hazard trends predicted by using the Bayesian smoothing technique for production months between July and November. All these plots show a relatively stable and constant trend after 2000 miles in service, although the predicted hazard rates appear to be at different levels for the five production months. In particular, the components manufactured in September, October and November appear to have higher hazard rates compared to July and August. This might be due to the fact that the estimates for these months are based on failure samples which are less consistent, as the vehicles from these production months have been less exposed to failure through mileage accumulation.

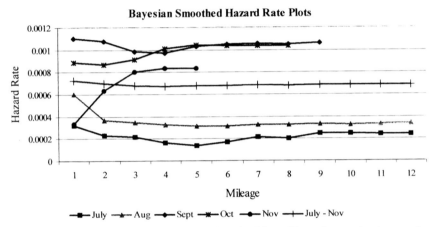

Figure 2. Bayesian smoothed hazard rate plots for July to November production months

Survival probability predictions can be obtained from Equation 11; Table 3 below gives reliability predictions based on combined failure data from July to November production months (the predictions for 36 and 100 k miles were made by assuming constant hazard rate pattern).

Table 3. Survival probability predictions (July – November data)

Mileage [k miles]	3	6	12	36	100
Survival Probability [%]	99.8	99.6	99.2	(97.6)	(93.4)

EXAMPLE 2 BATTERY WARRANTY ANALYSIS

Analysis of battery warranty failure data was required by an investigation aimed at correlating battery testing with field reliability performance. In particular, the comparative effect of customer usage profile from two different markets (UK and US, respectively) on battery failures was of interest. The warranty policy operated on both markets for the selected carline was 4 years in service (unconstrained mileage). Battery failure data for each market up to 40 MIS were extracted from the company's warranty database in the format presented in Table 1. Figure 3 shows comparative contour plots of the distribution of battery warranty failures in the two markets.

C606/018/2002 © IMechE 2002

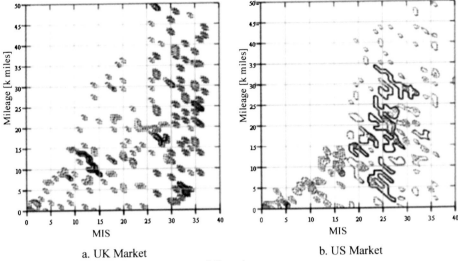

a. UK Market   b. US Market

Figure 3. Visualisation of battery warranty failure data

Battery time to failure analysis

Battery hazard rate analysis against time in service accumulation can be easily carried out using the warranty failure data presented in the format shown in Table 1. The MLE of the interval hazard rate can be calculated as follows:

$$h_i = \frac{\sum_j R_{i,j}}{0.5 \cdot \sum_j R_{i,j} + N_{i+1} - \sum_{q=1}^{i}\sum_j R_{q,j}} \tag{12}$$

Equation 12 assumes that censorings occur at exact times (at the end of a time interval), and failures occur at the middle of a time interval. The Bayesian smoothing algorithm was applied to the hazard rate function in Equation 12, and the resulting hazard rate plots for the two markets is shown in Figure 4. The plots show a bathtub like shape, with an initially decreasing hazard rate (up to 5 months in service) followed by a constant hazard rate pattern (up to 25 MIS for the UK market, and 20 MIS for the US market, respectively), and then an increasing hazard rate pattern towards the end of the observation period. Although the pattern is similar for the two markets, the actual hazard rate appears to be higher for the US market and the increasing hazard rate pattern occurs earlier in the battery life. Comparative reliability predictions are given in Table 4.

Table 4. Battery survival probability predictions (time in service)

MIS	3	6	12	24	36
UK [%]	99.8	99.7	99.5	99.1	98.4
US [%]	99.7	99.5	99.2	98.4	97.1

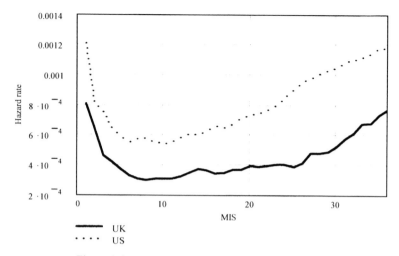

Figure 4 Comparative battery hazard rate plots (time to failure)

Battery mileage to failure analysis

To investigate the battery hazard rate pattern with mileage accumulation a lognormal distribution of the mileage accumulation process was used to calculate the expected number of censoring for each mileage interval. The MLE hazard rate for the j-th mileage interval can be calculated as follows:

$$h_j = \frac{\sum_i R_{i,j}}{500 \cdot \sum_i \left(R_{i,j} + c_{i,j}\right) + 1000 \cdot \sum_{q=j+1}^{M} \left(R_{i,q} + c_{i,q}\right)} \tag{13}$$

Equation 13 assumes that failures and censorings occur at the middle of a mileage interval (1000 miles). The Bayesian smoothing algorithm was applied to the hazard rate function in Equation 13, and the resulting hazard rate plots for the two markets is shown in Figure 5. The plots in Figure 5 show an initial decreasing hazard rate pattern followed by a constant (or slightly linearly increasing) hazard rate pattern. The patterns for the two markets are similar, but the hazard rate for the US market is higher than the UK's, meaning that more customers can be expected to experience battery failures during warranty in the US. This is also illustrated by the survival probability predictions given in Table 5.

Table 5. Battery survival probability predictions (mileage to failure)

Mileage [k miles]	5	10	20	30	40	50
UK [%]	99.5	99.1	98.3	97.4	96.4	95.3
US [%]	99.0	98.2	96.7	95.1	93.5	91.9

Given that Figure 7 shows that there is no significant increase in hazard rate with mileage it can be concluded that high mileage accumulators are not more likely to experience battery concerns,

C606/018/2002 © IMechE 2002

whereas battery aging increases the likelihood of failure (which is not unexpected). However, the hazard rate plots allowed a meaningful assessment in engineering terms of the hazard rate and patterns in the two markets. The fact that the hazard patterns are virtually identical indicates that the effect of customer usage on the failure mechanisms is similar. Engineering judgement based on understanding of the failure modes and mechanisms can be used to explain the differences between the hazard rates in the two markets. For example, it is known from the chemistry of battery operation that battery life correlates with environmental temperature. Therefore, environmental temperature is a factor that could explain the hazard rate difference, given that the US warranty data sample came from geographic area where the temperatures are considerably higher than in the UK.

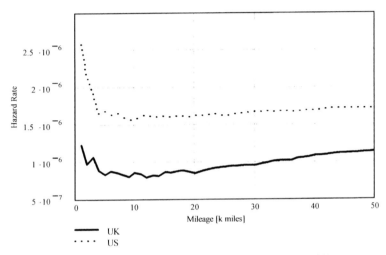

Figure 5. Comparative battery hazard rate plots (mileage to failure)

CONCLUSIONS

The paper demonstrated that hazard rate modelling of automobile warranty failure data is possible and has practical advantages. Hazard rate pattern analysis allows the engineer a good understanding of the reliability of the component under customer usage. Failure modes and mechanisms can be associated with trends in the hazard pattern (such as an early wear-out mechanism), which could point towards the appropriate reliability improvement actions. It is also a robust tool for decision concerning the reliability performance in the field.

It is interesting to note that Weibull analysis, a tool often chosen by reliability engineers, would not offer the same level of feedback. Moreover, Weibull analysis is more likely to fail in dealing with small and heavily censored samples and when a mixture in the failure data is present. From another point of view, the information obtained from the hazard rate analysis provides useful prior knowledge for fitting a statistical model (such as Weibull) to lifetime data.

ACKNOWLEDGEMENT

Research presented in this paper was partially supported by the UK Engineering and Physical Sciences Research Council, Grant GR/N06021.

REFERENCES

1. Lawless, J.F. 1982. *Statistical Models and Methods for Lifetime Data*. J Wiley & Sons.
2. Suzuki, K. 1985. Nonparametric Estimation of Lifetime Distributions From a Record of Failures and Follow-ups, *Journal of the American Statistical Association*, 80:389, 68-72.
3. Suzuki, K. 1985. Estimation of Lifetime Parameters From Incomplete Field Data, *Technometrics*, 27:3, 236-271.
4. Lawless, J.F., Hu, J. & Cao, J. 1995. Methods for the Estimation of Failure Distributions and Rates from Automobile Warranty Data. *Lifetime Data Analysis*, 1, 227-240.
5. Suzuki, K. 1993. Estimation Of Usage Lifetime Distribution From Calendar Lifetime Data. *JUSE Rep. Stat. Appl. Res.*; 40(1-2):10-22.
6. Hu, X.J, Lawless, J.F., Suzuki, K. 1998. Nonparametric Estimation of a Lifetime Distribution When Censoring Times are Missing, *Technometrics*, 40:1, 3-13.
7. Lawless, J.F. 1998. Statistical Analysis of Product Warranty Data, *International Statistical Review*, 66:1, 41-60.
8. Yang, G., Zaghati, Z. 2002. Two-Dimensional Reliability Modelling from Warranty Data. *2002 Proceedings Reliability & Maintainability Symposium*, 272-278.
9. Wasserman, G.S., Sudjianto, A. 1996. A Comparison Of Three Strategies For Forecasting Warranty Claims. *IIE Transactions*, 28, 967-977.
10. Lu, M.W. 1998. Automotive Reliability Prediction Based on Early Field Failure Warranty Data. *Quality and Reliability Engineering International*, 14, 103-108.
11. Campean, I.F., Kuhn, F. & Khan, M.K. 2001. Reliability Analysis of Automotive Field Failure Data, in Zio, Demichela & Piccinini (Editors) *Safety and Reliability*, Proceedings of ESREL 2001, 1337-1344.
12. Davis, T. 1999. A Simple Method for Estimating the Joint Failure Time and failure Mileage distribution from automobile warranty data. *Ford Technical Journal*, vol. 2, issue 6.
13. Gammerman, D. 1994. Bayes Estimation of the Piecewise Exponential Distribution. *IEEE Transactions on Reliability*, 43(1):128-131.
14. Campean, I.F. 2000. Exponential Age Sensitive Method for Failure Rank Estimation. *Quality & Reliability Engineering International*, 16:291-300.
15. Kim, J.S., Proschan, F. 1991. Piecewise Exponential Estimator of Survivor Function. *IEEE Transactions on Reliability*, 40:2, 134-139.
16. Kaplan, E.L. and Meier, P. 1958. Nonparametric Estimation From Incomplete Observation. *Journal of American Statistical Association*, 53, 457-481.

C606/018/2002 © IMechE 2002

C606/008/2002

Application of proportional hazard model to tyre design analysis

V V KRIVTSOV, D E TANANKO, and **T P DAVIS**
Ford Motor Company, Dearborn, Michigan, USA

ABSTRACT

This paper considers an empirical approach to the root-cause analysis of a certain kind of automobile tire failure. Tire life data are obtained from a laboratory test, which is developed to duplicate field failures. A number of parameters related to tire geometry and physical properties are selected as explanatory variables that potentially affect a tire's life on test. Analysis of the survival data is performed via the proportional hazards model. The obtained statistical model helps to identify the elements of tire design affecting the probability of tire failure due to the failure mode in question.

1 INTRODUCTION

By the design intention, an automobile tire should exhibit no failures during its useful life and/or while the tread depth is still adequate. However, some tires do fail prematurely. There are several kinds of failure modes observed in the field. This paper focuses on a particular failure mode known as *tread and belt separation (TBS)*. In the event of TBS, the whole (or a part of the) tread and the second (upper) steel belt leave the tire carcass and the first (lower) steel belt (see Figure 1).

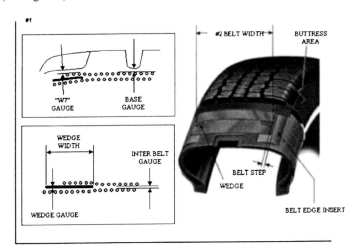

Figure 1. Elements of Radial Tire

Usually, this failure occurs at highway speeds, so pieces of tread produce local damage to the vehicle body. More importantly, the rubber between steel belts (whose durability characteristics differ from those of tread material) becomes exposed to the road surface. This failure can affect lateral stability of the vehicle.

2 ENGINEERING HYPOTHESIS

One could consider TBS as a sequence of two events: failure crack initiation in the wedge area (which usually starts as a "pocketing" at the edge of the second belt) followed by the crack propagation between the belts. Finite element analysis of tire geometry (1) suggests that the largest strain occurs in the wedge area and is proportional to the wedge gauge. Moreover, the location of the wedge is critical to heat dissipation.

While a small wedge encourages crack initiation, it is not itself a sufficient condition for TBS to occur. In order to propagate further, the crack must have favorable conditions, e.g., low adhesion strength between belts and the proper energy input to separate the belts. These characteristics depend on physical, chemical, and mechanical properties of the rubber skim stock as well as the age of the tire. Hence, the following tire design characteristics have been selected as *explanatory variables (covariates)* that could potentially affect the tire's life until the TBS failure:

- Tire age
- Wedge gauge
- Interbelt gauge
- End of belt #2 to buttress
- Peel force (adhesion force of rubber between steel belts, characterized as the force required to separate belts in the specimen of a given dimension)
- Percent of carbon black (a chemical ingredient of the rubber affecting its mechanical characteristics, such as *tear resistance*)

Field failure data turn out to be insufficient for the construction of a tire reliability model with explanatory variables. While the survival times *can* be estimated and even censoring can be properly accounted for (2), the data on the above-defined covariates are difficult to obtain because of the disintegration of the tire as a result of TBS. In order to overcome this problem and duplicate field failures in controlled conditions, a special laboratory test has been developed.

3 TESTING PROCEDURE AND FAILURE WARNING SYSTEM

The laboratory study was performed on a mixture of new and field-exposed 15-inch radial tires manufactured at different plants. The testing was conducted on a dynamometer drum with monotonically increasing speed steps, at 100 degrees F, under the inflation pressure of 26 PSI. Because of high variability in tire life, the testing procedure involved loads of 1300 lbs and 1500 lbs; that is, for the tires that did not fail under the lower load, the higher load was applied. The test procedure consisted of three parts (see Figure 2):

- Warm up over 2 hours at 50 mph
- Cool down over 2 hours at full stop

C606/008/2002 © IMechE 2002

- In the 1300 lbs regime: speed steps starting at 75 mph and increasing by 5 mph every half hour till 90 mph and then every hour till failure
- In the 1500 lbs regime: all the above speed steps are of half-hour duration

The test procedure above is a modification of the *high-speed drum test*, which is widely used in the tire industry as an accelerated key life test to identify potential failure modes, compare different designs, and validate design changes.

Several different failure modes can be observed during this test. Some of them are accompanied by detachment of the large chunks of material and even full disassembling of the tread and/or belt(s).

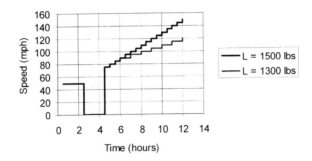

Figure 2. Test Speed Profiles

To overcome the contradiction between the destructive nature of the test and the need to properly measure geometry and material properties of the tire, a special failure warning system has been developed (3).

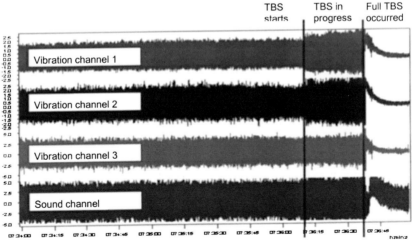

Figure 3. Vibration and Sound Pattern of Tire before TBS Event

The main idea of this system is based on the fact that the internal crack must be developed inside the tire prior to its catastrophic failure. As a result of centrifugal forces at high speed, the large chunk of material lifts up at the crack location, which accompanied by further crack growth, then leads to a change of the tire eccentricity. Therefore, the vibration signature of the rotating tire can be used for early detection of the failure.

The failure warning system involves specially mounted accelerometers and a PC-based data acquisition system. A signal to stop the test is generated when the system senses the change in the tire's vibration pattern (at the first vertical line in Figure 3). If the test is not stopped, it leads to a full TBS separation. The associated time to *partial* TBS is thus equalized with that of *full* TBS for data analysis purposes. The partially disintegrated tire is then conveniently available for further tests to properly measure the mechanical, physical, and chemical covariates.

4 STATISTICAL MODEL

The proportional hazard model (4) offers a convenient and physically meaningful way of relating the life characteristic of an item to the vector of explanatory variables. According to this model:

$$h(t,z) = h_0(t) \cdot exp(\beta^T \cdot z)$$

where: t is the time to failure, $h(t,z)$ is the hazard rate, contingent on a particular covariate vector (of explanatory variables) z, $h_0(t)$ is the baseline hazard rate (when all explanatory variables are equal to zero), β^T is the transposed vector of regression coefficients.

The advantage of the Cox model over parametric survival regression models is that it does not make any assumption about the nature or the shape of the underlying survival distribution, thus reducing the uncertainty about model selection. The statistical estimation of the Cox model parameters is possible through maximization of the simplified partial likelihood function (5).

All covariates identified in Section 2 and included in the model have been checked for statistical independence and lack of autocorrelation. Failure times associated with competing failure modes (other than TBS) have been treated as censored responses.

5 DATA ANALYSIS

In order to account for the difference in speed profiles between the two loads cases, the survival variable has been transformed from *time to failure (TTF)* to equivalent *virtual work* done against the tire until the failure, that is:

$$W = L * S,$$

where L is load against the tire, and S is the mileage passed by the tire on the test.

It must be noted that this is not *actual* work done against the tire due to the rolling resistance, but a cumulative characteristic proportional to the applied load and mileage of the tire until failure.

C606/008/2002 © IMechE 2002

Table 1. Test Data Set Used in Proportional Hazard Analysis

Tire Age	Wedge Gauge	Interbelt Gauge	EB2B	Peel Force	% Carbon Black	Wedge Gauge x Peel Force	Survival	Censoring (1-compl, 0-cens)
1.22	0.81	0.88	1.07	0.63	1.02	0.46	1.02	0
1.19	0.69	0.77	0.92	0.68	1.02	0.43	1.05	1
0.93	0.77	1.01	1.11	0.72	0.99	0.49	1.22	0
0.85	0.80	0.57	0.98	0.75	1.00	0.42	1.17	1
0.85	0.85	1.26	1.03	0.70	1.02	0.64	1.09	0
0.91	0.89	0.94	1.00	0.77	1.03	0.59	1.09	1
0.93	0.98	0.84	0.92	0.72	1.00	0.55	1.17	1
1.10	0.76	0.94	1.01	0.84	0.98	0.55	1.10	0
0.95	0.53	0.96	0.91	0.58	1.00	0.27	1.00	1
0.94	0.87	1.11	0.88	0.72	0.99	0.65	1.15	1
1.08	1.13	1.12	0.93	0.75	0.96	0.79	0.98	1
0.89	1.03	1.28	0.97	0.68	1.02	0.53	1.24	0
1.41	0.79	0.83	0.91	1.00	1.00	1.00	0.98	1
1.50	0.72	0.76	0.97	0.76	0.96	0.35	1.15	1
1.21	0.54	0.70	0.95	0.59	1.00	0.30	0.65	1
2.01	0.76	0.94	1.01	0.53	1.00	0.35	0.97	1
1.49	0.64	0.70	1.02	0.71	0.97	0.41	0.85	0
1.55	0.63	0.71	1.13	0.66	1.00	0.40	0.98	0
1.23	0.84	1.09	1.04	0.76	0.98	0.57	1.02	0
2.60	1.05	1.21	1.07	1.06	0.99	1.05	1.14	0
2.26	0.98	1.34	1.02	0.87	1.00	0.89	1.18	0
1.66	1.13	0.68	1.18	1.02	0.98	0.86	1.18	0
2.03	0.96	1.12	1.11	0.57	1.01	0.47	0.91	0
0.38	1.15	1.01	0.97	0.81	1.00	0.86	0.75	0
0.45	1.23	1.01	0.96	0.74	1.00	0.91	0.79	0
0.38	0.89	1.03	0.99	0.84	0.99	0.74	0.87	0
0.09	1.37	1.29	1.06	2.27	1.00	2.61	0.87	0
0.09	1.35	1.44	0.95	2.33	1.00	3.00	0.87	0
0.09	1.49	1.13	0.91	2.15	1.00	2.75	0.90	0
0.15	1.32	1.11	0.91	1.90	1.00	2.18	0.91	0
0.17	1.68	1.12	1.05	1.74	1.02	2.44	0.79	0
0.17	1.71	0.98	1.05	1.68	1.02	2.42	0.83	0
0.17	1.63	1.05	1.02	1.44	1.03	2.16	0.84	0
1.05	1.04	1.06	1.02	1.03	1.03	0.93	1.28	0

Table 1 shows the test data set (coded for confidentiality). The results of the survival regression analysis are shown in Table 2. The log-likelihood of the final solution is -16.008, while the log-likelihood of the null model (with all regression parameters being equal to zero) is -28.886. The likelihood ratio chi-square statistic (the null model minus the final solution) is 25.757 with 7 degrees of freedom and the associated p-value is 0.0005. Highlighted covariates are statistically significant at $p < 0.05$.

The developed life test does not appear to be sensitive enough to distinguish between aged and new tires, hence the non-significance of the tire age covariate. See Baldwin (6) for a more detailed discussion of the tire age factor.

Table 2. Estimates of Proportional Hazard Model with all Covariates

Explanatory Variable	Beta	Standard Error	t-value	p-value
Tire age	2.109	1.393	1.514	0.130
Wedge gauge	**-9.686**	**4.638**	**-2.088**	**0.037**
Interbelt gauge	**-10.677**	**4.617**	**-2.313**	**0.021**
Belt2 to sidewall	-13.675	8.112	-1.686	0.092
Peel force	**-34.293**	**13.651**	**-2.512**	**0.012**
% Carbon Black	-48.349	33.448	-1.445	0.148
Wedge x Peel force	**20.839**	**8.860**	**2.352**	**0.019**

Table 3 shows the estimation results of the model that includes only statistically significant covariates. The log-likelihood of the final solution is -19.968, while the log-likelihood of the null model is -28.886. The likelihood ratio chi-square statistic is 17.837 with 4 degrees of freedom and the associated p-value is 0.001.

Table 3. Estimates of Proportional Hazard Model with Statistically Significant Covariates

Explanatory Variable	Beta	Standard Error	t-value	p-value
Wedge gauge	-9.313	4.069	-2.289	0.022
Interbelt gauge	-7.069	2.867	-2.466	0.014
Peel force	-27.411	10.578	-2.591	0.010
Wedge A x Peel force	18.105	7.057	2.566	0.010

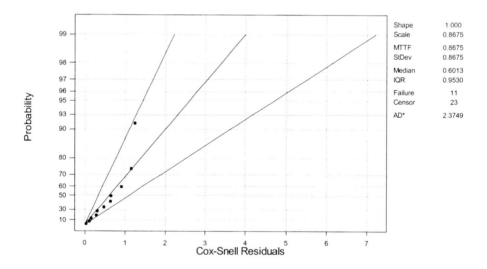

Figure 4. Exponential Probability Plot of Cox-Snell Residuals

C606/008/2002 © IMechE 2002

Shown in Figure 4 is the exponential probability plot of Cox-Snell residuals, which confirms the adequacy of the fitted model.

Figure 5 displays the cumulative hazard function (characterizing tire's propensity to failure) predicted from the estimated model based on some typical values of covariates for "poor" and "good" tires. The poor tire is assumed to have wedge and interbelt gauges of 0.5 and peel force of 1, and the good tire, wedge and interbelt gauges of 1.2 and 1, respectively, and peel force of 2.

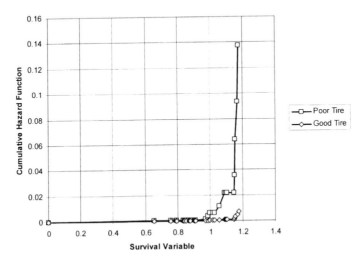

Figure 5. Cumulative Hazard Function Predicted from the Estimated Model Based on Some Typical Values of Covariates for "Poor" and "Good" Tires

6 CONCLUDING REMARKS

The statistical analysis of laboratory test data shows that the wedge and interbelt gauges as well as the peel force are significant factors affecting the hazard rate of TBS failures in an inversely proportional way. This is in good agreement with the engineering hypothesis formulated above. The obtained results should be viewed as qualitative (i.e., helping to compare tire designs from a reliability standpoint) rather than quantitative (i.e., predicting the actual reliability of a tire in the field).

REFERENCES

1. DeEskinazi, J., Ishihara, K., Volk, H. and Warholic, T.C. (1990) Towards Predicting Relative Belt Edge Endurance with the Finite Element Method. *Tire Science and Technology*, Vol. 18, 4, 216-35.

2. Davis, T.P. & Krivtsov, V.V. (2002) The Role of Statistical Science in Firestone Tire Failure Root Cause Investigation. Work in progress to appear in *Journal of Quality Technology*.

3. Tananko, D.E. (2002) A Nondestructive Monitoring and Alarm System for High Speed Tire Test. Ford patent pending.

4. Cox, D.R. (1972) Regression Models and Life Tables (with discussion). *J. R. Statist. Soc. B*, 34, 187-202.

5. Cox, D.R. (1975) Partial Likelihood. *Biometrica*, 62, 269-76.

6. Baldwin, J. M. (2002) Correlation of Physical And Chemical Properties of Artificially Aged Tires Vs. Field Aged Tires - Part 1. To appear in Proc. *Spring Symposium of the American Chemical Society - Rubber Division*, Savannah, Georgia. April, 2002.

      C606/008/2002 © IMechE 2002

C606/009/2002

Better understanding of automotive manufacturing reliability data using a spreadsheet-based model fitting tool

M LINSLEY, T FOUWEATHER, D McGEENEY, D J STEWARDSON, and S Y COLEMAN
Industrial Statistics Research Unit, University of Newcastle-upon-Tyne, UK

ABSTRACT

This paper describes the nature and use of a pair of simple spreadsheet based model fitting tools for use with reliability type data. These can be used in a number of applications, and were developed during work on improvements for a large North Eastern car manufacturer. The tool is based on the use of probability plots, these being visual methods of assessing the fit of data to a number of potential statistical models. The tools enable simple adjustment of the data if required, conducted along valid statistical lines, as protection against unusual early results. We illustrate the models with data from the car industry.

1. INTRODUCTION

The need for better understanding about the expected useful lifetime of products has followed the introduction of new requirements for 'whole-life-cycle' planning and costing. Warranty databases are being set up and analysed all over Europe. Producers want to know if they may face future claims, in particular where warranties cover long periods of time. Consumers now expect products, especially vehicles, to last longer than they have ever lasted before. This paper presents a simple way of analysing lifetime or time-to-event data whether this is from in-house accelerated testing, or from the field. This is based on (a) spreadsheet based probability plots, and (b) a Weibull probability distribution fitting tool, also in a spreadsheet. Engineers can use the first to establish which of several possible statistical 'models' best fit their data, and the second to actually fit a model that helps to explain the nature of that data. This then produces such supplementary statistics as the mean time to failure, the failure or hazard rate, the survivor functions and curves and plots to help envisage all of these. The tools are easy to use, require no previous advanced statistical knowledge and enable reliable estimates of various reliability measures to be obtained for a variety of time-to-event data types.

2. PROBABILITY PLOTS

These are one of the most useful statistical tools available. Past applications have tended to rely on the use of graph paper, as described in texts such as King (1). These graph papers were set up such that the axes incorporated a transformation dependent on the distribution that was suspected and being tested for. Modern use has tended to rely on the computer and there have been many uses for the plots demonstrated over the years. Daniel (2) introduced their routine use in the analysis of designed experiments, in particular the Half-Normal plot version, and all modern statistical packages include Normal probability plots in their diagnostic toolbox. These methods are available for use with any distribution, not just the Normal, and some texts have demonstrated the transformations that allow their use in spreadsheet form, e.g. Kolarik (3) Kennedy & Neville (4). Despite this, even as late as the 1970s some writers seemed unaware of the possibilities, such as Gerson (5) who suggested, quite wrongly, that the chi squared distribution could not be graphed effectively in this way.

2.1 Spreadsheet Form

The paper and pen versions of these plots were sometimes difficult, and time-consuming and this may account for their relative unpopularity. However developments in modern spreadsheets has enabled a revival of the use of these simple but effective tools. The main difference in spreadsheet form is that the transformation is not based on the graph itself, but rather by way of formulas embedded in the spreadsheet. These vary by distribution, but the effect is always the same. A good fitting plot will produce a straight line roughly from corner to corner of the plot, see for example Hines and Montgomery (6).

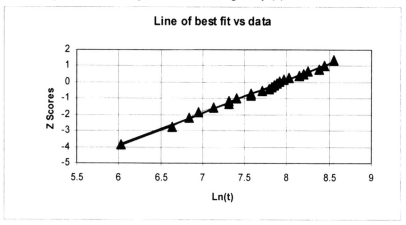

Figure 1 Weibull probability plot and 'fitted line' from spreadsheet.

Figure 1 shows a Weibull probability plot for an auto manufacturers reliability data, with a superimposed Regression line through the data. (For brief explanations of Ln(t) and Z see section 3.3.1). The straight line seen here is typical of a 'good' fit of the data to a 'model' in this case a Weibull distribution model. The tool that we use and describe here fits this, plus other statistical models, such as; Normal, Log-Normal, Half Normal, Extreme value, Cauchy, Gamma, Logistic, Uniform, Chi-square and the Exponential. The specific transformations for the several models available in this form are given in either Kolarik (3) or Kennedy & Neville (4).

An example of the fit of the data shown in Weibull form above but fitted to other potential models follows in Figures 2 and 3, being the Logistic and Exponential models respectively.

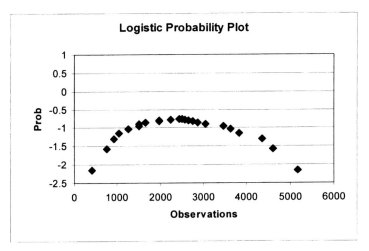

Figure 2 Logistic probability plot for reliability data.

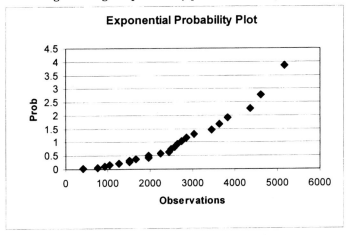

Figure 3 Exponential probability plot for reliability data.

It can easily be seen that a visual inspection of the plots is all that is required to establish that these models would be insufficient as a fit to this data. By looking at all the plots in the spreadsheet, the best model can be found by inspection. Some data will fit more than one model, for example a Normal fit will usually also fit the Log-normal and the Weibull as shown in King (1).

3 THE WEIBULL MODEL

The Weibull was named after its Swedish inventor and is well described in Weibull (7).

The Weibull model has 3 parameters, which are:

α - The scale parameter, which controls the amount of spreading from left to right evident in the graph of the probability density function (pdf).

β - The shape parameter that determines the basic form of the pdf

γ - The location parameter, which establishes the position of the start of the left end of the distribution, the point on the x-axis at which the pdf begins.

The probability density function (pdf) for the Weibull distribution is given by:

$$f(t) = \frac{\beta}{\alpha}\left(\frac{t-\gamma}{\alpha}\right)^{\beta-1} e^{-\left(\frac{t-\gamma}{\alpha}\right)^{\beta}}$$

where t must be $> \gamma$.

The cumulative distribution function (cdf) is:

$$F(t) = 1 - e^{-\left(\frac{t-\gamma}{\alpha}\right)^{\beta}}$$

All the functions used to describe the properties of the process or products being modelled are derived from these basic functions and the parameters that determine them. F(t) is the failure rate function that describes the cumulative proportion of failed items at any point in time. The reliability function R(t) = 1- F(t) represents the reliability rate (or survival) function and f(t)/R(t) that is the pdf divided by the survival function, produces the hazard function that describes the instantaneous probability of failure for a product at any point in time.

These represent the most useful of the functions in practice. Several modern statistical computer packages have Weibull distribution outputs but these often require user input of the distribution parameters, for example Minitab version 12 (8) offers no way of finding the parameters if they are unknown. This paper shows an easy way of determining them using a spreadsheet.

3.1 Model Fitting

The Weibull, is one of the most versatile probability distributions available. Unlike the commonly used and simpler Exponential model, (that assumes a constant failure rate throughout) see for example Bentley (9) the Weibull can handle changing failure rates. This is true whether the rate is an increasing or decreasing one. It is also particularly important because it allows the user to adjust the data according to the average starting 'time' of failures (or other events such as returns), in order to better fit model to the data. This is known as 'correcting for Gamma' where gamma is one of the parameters in the model, the one concerned with the 'starting time' for the event being modelled. This concept can be confusing, so a fuller non-mathematical explanation may be useful. The time to early failures of a product or component are subject to random variation, as in any probability situation. So a good fit in this model requires an estimate of the average or expected first failure time, in effect, an adjustment from any very early failures due to these random effects. Occasionally a backwards adjustment is needed, where there is no failure seen early enough (again due to chance) but this correction is usually forwards in time.

3.2 Correcting for gamma

The Weibull distribution is effectively reduced to a two-parameter distribution by making this initial correction in the data for the time parameter γ, by subtracting the value for the estimated or expected time to first failure from all subsequent failure times. This is required to establish the correct shape and scale parameters for the hazard, reliability and failure rate functions. When a changing failure rate exists, as it often does, the Weibull probability plot is only straight if the t values are corrected for the estimated first time to failure. In the old graphical procedure that was used to find a good fitting Weibull model, as King (1) and Capelin (10) show, the gamma correction was estimated using a probability plot and only part of the original data. A revised probability plot, using the gamma-corrected data, was then used to find the corrected model parameters. However this procedure allowed the possibility that the final model was not the best possible fit to the data. Reliance on just part of the original data is rarely best practice. The method shown here fits the data directly by iteratively correcting the original data for time to first failure until a best probability plot fit is achieved. We now look at the proposed procedure. A comparison with the older graphical method was given in Burdon et al. (11).

3.3 Spreadsheet procedure

3.3.1 Initial check to see if data follows a Weibull distribution

Most of the data considered here involves the times to failure of particular items.

The initial stage is to establish if the Weibull is a reasonable model for the data being investigated. Using the spreadsheet of probability plots, the Weibull version is a plot of Ln(t) against a derived 'Z' value as defined below, the data is plotted. Ln(t) is the natural logarithm of the original data; t. Often a transformation of the data needs to be performed before it can be fitted to a distribution. In this case the Ln(t) transformation is appropriate. During this first stage there will have been no 'correction for gamma'. The Z values are calculated as $Ln(-Ln(1-F(t)^*))$ where $F(t)^*$ is the empirical estimate of the failure rate function (the un-reliability) of the product under investigation and is simply the empirical estimate of the probability that some product has failed to meet the specifications for use by time t.

$F(t)^*$ is calculated as $(n_t - c)/N$ where n is the number of failures up to a given time t, c is a continuity correction often set at 0.5, see Chatfield (12), and N is the total number of items that could fail, usually the total production in warranty data cases. From the formula for F(t) above it can be seen that one way to investigate the data is to plot $Z = Ln(-Ln(1-F(t)))$ against $Ln(t-\gamma)$. At this stage γ is not guessed at and the unadjusted values of t can be used. (i.e. plot Z versus log(t)).

With regard to the continuity correction, other valid methods include the probability plot correction $F(t) = (3t-1)/(3N+1)$ as given by Tukey (13) or the version used by Minitab $F(t) = (t-3/8)/(N+1/4)$ introduced by Blom (14). As discussed by Chatfield (12) the choice makes little difference in practice. To test whether or not the data follows the chosen distribution, i.e. the Weibull distribution; the transformed data must be plotted against standard values. For the Weibull distribution, the standard values are $Z = Ln(-Ln(1-F(t)))$. If the transformed data plotted against the standard values approximates to a straight line then the Weibull distribution will probably produce a good fit to the data. Further details of the calculation of Z and its derivation, with a proof that its relationship with Ln(t) is equivalent to the original Weibull probability paper based plots is given in Bentley (9).

3.3.2 Example using a spreadsheet
The following data represents the number of weeks from sale until a failure in a certain component is reported for a particular make of domestic car. This is typical of the kind of in-field warranty data that often occurs in practice.

Of 53000 cars these are the only 12 component failures that had been reported:

85, 41, 66, 51, 44, 58, 72, 102, 116, 108, 95 and 151 weeks from sale of vehicle.

The data is placed in ascending order, as shown in table 1, and F(t)* the empirical failure-rate function or un-reliability, and the associated Z values are calculated. Recall that R(t)* is simply 1-F(t)* and is the empirical Reliability function based directly on the data. The statistical model based versions of these functions will be denoted F(t) and R(t).

Table 1 example of initial warranty data being fit to Weibull using spreadsheet

Corrected Data (t-40)	Fails	Raw Data (t)	F(t)*	R(t)*	Ln(t)	Z
1	1	41	9.434E-06	0.99999	0	-11.571
4	2	44	2.83E-05	0.99997	1.38629	-10.473
11	3	51	4.717E-05	0.99995	2.3979	-9.9617
18	4	58	6.604E-05	0.99993	2.89037	-9.6253
26	5	66	8.491E-05	0.99992	3.2581	-9.3739
32	6	72	0.0001038	0.9999	3.46574	-9.1732
45	7	85	0.0001226	0.99988	3.80666	-9.0062
55	8	95	0.0001415	0.99986	4.00733	-8.8631
62	9	102	0.0001604	0.99984	4.12713	-8.7379
68	10	108	0.0001792	0.99982	4.21951	-8.6267
76	11	116	0.0001981	0.9998	4.33073	-8.5266
111	12	151	0.000217	0.99978	4.70953	-8.4356

In Table 1 the total number of cars manufactured in this period (N) was 53000. Therefore, the un-reliability F(t)* in column 4 is calculated by (n-c)/53000. For example using the last row in the table, when n =12 (column 2), and c is set at 0.5, then F(t)* =(12-0.5)/53000 = 0.0002170 (column 4). The Z value (column 7) is calculated as $Z = ln(-ln(1-F(t)^*))$. For example, when n=12 the Z value is -8.436 as shown in the table.

The corresponding probability plot of Z against ln(t) is shown in Figure 4. The plot suggests an approximate straight line indicating that the data will probably fit a Weibull distribution based model once it has been corrected for gamma. This is the next step in the process.

3.3.3 Correcting for Gamma

The next stage of the process is the Correction for Gamma. The initial attempt to correct for gamma is made by subtracting an estimate of the time to first failure from all the failure times in the data. In this example we start by subtracting 40 weeks from all failure times. The observed time to the first failure is 41 weeks so this means that the corrected first data point is now 1 week. Figure 5 shows the new probability plot after the correction.

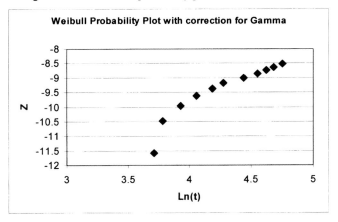

Figure 4 plot of Z Vs Ln(t) for example data

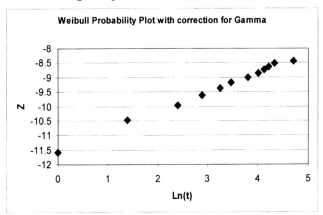

Figure 5 plot after correction for gamma of 40 weeks

It is possible to fit a regression line to this relationship, with Ln(t) as the predictor variable and Z as the response. The best fit to the data is determined by minimising the sum of squares of residuals (the differences between the corresponding observed values and the regression predictions) for the corrected data. Continuing with the example, the results of the correction for gamma of 40 produces a sum of squared residuals of 0.0429. Figure 1 showed an example of such a fit. We can now try adjusting the correction value until we get a best-fit regression. We adjust the correction until the minimum sum of squared residuals is achieved. In this example when we reduce the correction for gamma to 39, this produces an inferior plot and an increase in the sum of squares to 0.1115. With a correction of 38 the sum of squares increase to 0.1973. Clearly 40 is the 'best' correction.

From this quick and easy use of spreadsheets we have determined an optimum correction and we can now establish that the Weibull distribution parameters are:

β = 0.6719. which is simply the slope of the regression line and α = 2817 calculated by (-d/β) where d, is the intercept of the regression line (9).

3.3.4 Hazard rate

When using the model to determine hazard rates and cumulative failures against time we must remember to re-adjust the time to failure and associated predicted times by the 40 week correction required to find the best fitting model. In other words we must add the 40 weeks back to any given estimate of elapsed time to failure.

Note that the value of beta (less than 1) is indicative of a falling failure rate. Plotting the hazard function best shows this. Recall that this is a function of R(t) and the pdf.

$$\frac{pdf}{R(t)} = \frac{\beta}{\eta}\left(\frac{t-\gamma}{\eta}\right)^{\beta-1}$$

For this example the hazard function looks like figure 6

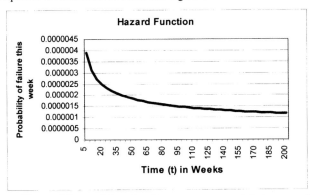

Figure 6 Hazard function for car component data example

The plot shows that the probability of a failure has fallen and continues to fall over time. The higher failure rates are typical of early failure rates seen in electronic component failures.

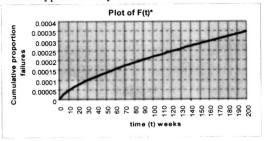

Figure 7 plot of cumulative failures

C606/009/2002 © IMechE 2002

The cumulative failure rate is given by F(t). A plot of F(t) (figure 7) shows how many failures we can expect over time. At each time the proportion of failures is indicated. From this plot we can establish how many failures to expect by a given time. For example, if our warranty period is 3 years = 156 weeks, we can see from the plot that by this time (156-40 = 116 allowing for the correction) we can expect 0.000241 of the total of the components to have failed, and this represents a total of 53000(0.000241) = 12.8 or 13 rounded to a whole number of failures for this component over a three year warranty period.

Naturally with sparse information (12 failures only so far) this estimate is subject to some uncertainty but does represent the best estimate at this stage.

3.3.5 Mean time to failure
Note that we can also derive an estimate of the Mean Time To Failure, that is often quoted in reliability reports, using the Excel Gamma function (Γ) via the formula

$$MTTF = \gamma + \alpha\Gamma\left(1 + \frac{1}{\beta}\right)$$

In this case the MTTF is very long, many thousands of years, indicating that we have a very reliable component. Of course in reality the component would wear out long before this but based on the current data, and without wear-out, the component would easily outlast the car.

4. CONCLUSIONS

The benefits of using these spreadsheet tools are:

- They are simple to use and require no expert statistical knowledge
- The graphical approach helps with ease of useA number of distributions are covered, some rarely used
- They provide an easy means by which to convey the findings to managers
- The Weibull fitting tool provides 'best-fit' model parameter estimates.
- Correction for early time to failures is available
- Reliable estimates of reliability parameters can be established.
- The ease with which the formulae behind the calculations can be traced using a spreadsheet can help trainees understand where the data is linked and why certain calculations are performed.
- The graphs can be used to provide good estimates of failure rates.
- A 'black-box' approach is avoided.

5. THE SPREADSHEET

The spreadsheet set is available for general use, by request to the authors, with accompanying user instructions.

REFERENCES:

(1) King J.R. (1971) **Probability Charts for Decision Making** Industrial Press New York

(2) Daniel C (1959) Use of Half-Normal Plots in Interpreting Factorial Two-Level Experiments **Technometrics vol 1 (4)**

(3) Kolarik W J (1995) **Creating Quality**, McGraw Hill New York

(4) Kennedy J B, Neville A M, (1976) 2nd ed. **Basic Statistical Methods for Engineers and Scientists** Harper & Row, New York

(5) Gerson M (1975) The Techniques and Uses of Probability Plotting **The Statistician Vol 24 (4)**

(6) Hines WH & Montgomery DC (1990) **Probability and Statistics in Engineering and Management Science** Wiley, New York

(7) Weibull W (1951) Statistical Distribution Function of Wide Application **Journal of Applied Mechanics v18 p293**

(8) Minitab release 12 (1997) Minitab Incorporated USA

(9) Bentley J.P. (1999) 2nd ed. **Introduction to Reliability & Quality Engineering** Addison Wesley Longman Harlow

(10) Caplen R.H. (1972) **A Practical Approach to Reliability** Business Books London

(11) Burdon C, Fouweather A, Greaves LH, Stewardson DJ (2000) **The use of Spreadsheets to fit Weibull Distribution based Models with Correction for Gamma** Proceedings of the International Conference on Industrial Statistics In Action, Newcastle

(12) Chatfield C (1995) 2nd ed. **Problem Solving A Statisticians Guide** Chapman Hall, Boca Raton

(13) Tukey (1962) The Future of data Analysis **Annals of Mathematical Statistics v33 p1**

(14) Blom G (1958) **Statistical Estimates and Transformed Beta Variables** Wiley, New York

C606/009/2002 © IMechE 2002

C606/012/2002

Achieving robust designs through the use of 'design for Six Sigma'

D MORTIMER
Product Design Group, Leyland Trucks Limited, Preston, UK

SYNOPSIS

While conventional Six Sigma techniques have proved successful in most departments at Leyland Trucks, we have struggled to apply the methodology to the majority of the work we undertake in the Product Design Group (PDG). We have now adopted 'Design for Six Sigma' (DFSS) techniques and we find that this revised methodology is much better suited to our needs.

We believe that the use of DFSS methods will result in the implementation of robust designs. In turn, we believe that these robust designs will reduce development lead times and improve Right First Time (RFT) performance. Through the use of case studies we hope to prove both these assertions.

1   INTRODUCTION

Conventional Six Sigma follows the Define, Measure, Analyse, Improve, Control (DMAIC) methodology. This has proven to be well suited to problem solving and process improvement projects. We assert, however, that it is not well suited to improving product design. In contrast, DFSS uses the Identify, Define, Design, Optimise, Validate, Evaluate (ID DOVE) methodology. These steps fit neatly with our Product Creation Process (PCP) framework of Definition, Concept, Engineering, Development, Validation, Introduction (1) (See Figure 1).

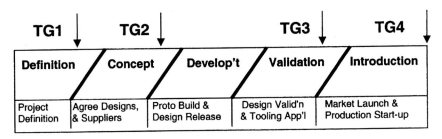

Figure 1.

The fact that DFSS methodology fits so neatly into the framework we already have in place, allows us to promote the use of inferential statistics and thereby an understanding of the inherent variability in production processes. This in turn allows us to promote robust design.

Implementation of the DFSS methodology requires that the purpose of the project be suitably defined before the product requirements are identified. Several design concepts are then proposed and suitable decision making processes are used to select the option that is most appropriate. Having chosen which design to use, this is then optimised. Often it is at this stage that it is appropriate to develop an understanding of the inherent variability of the processes involved. This is usually achieved by calculating the mean and standard deviation of either the suppliers process or our own. From this understanding of the capability of the manufacturing process our engineers are able to choose suitable tolerances. Additionally, from understanding this capability the engineer is better able to assess whether statistical tolerancing is appropriate. These are some of the key steps in the process. Before the design is introduced validation must take place to ensure that the design meets the requirements that were identified earlier and following introduction, the impact of the new design is evaluated and documented.

2   COMPARISON: CONVENTIONAL SIX SIGMA TO DFSS

2.1   Conventional Six Sigma
The following steps are taken in a conventional Six Sigma project.

2.1.1   Define
The format of a Six Sigma 'problem statement' is rigidly defined. It should be phrased in the following format: "To reduce (or increase)......a measurable entity.....from x to y" (where x is the current measurement and y is the goal).

The problem statement naturally leads to a 'Defect Definition' of the following form: "Any......measured entity.....greater (or less) than y".

The following is a typical example:
Problem Statement: "Reduce Engineering Change Order (ECO) introduction time from two months to two weeks."
Defect Definition: "Any ECO introduction which takes longer than two weeks"

The 'belt' (Six Sigma practitioner) then defines the team, which should include the product or process owner as well as representatives from all the areas involved. A brainstorming session then takes place with the team where the process is defined. All the steps in the process are recorded in a process map which may take the form of a "Process Flowchart" or an "Input Process Output" (IPO) map. This is used to identify all possible factor that may influence the output detailed in the problem statement (e.g. all factors that might influence ECO introduction time). This may also be supported by the use of an Ishikawa Diagram (also referred to as a fishbone diagram).

2.1.2   Measure
Measurements are then taken to define the current performance level. In the above example there may already be a process in place that records the time taken for ECO introductions. If

not, the problem statement may exist for a time with "from x to y" included. The figures being inserted as and when determined. If a measurement process already exists it must be subjected to a "Measurement System Analysis" (MSA) and if not initially suitable it must be improved. If no measurement is in place a measurement system must be devised and an MSA must then be carried out. The MSA may be nothing more than a logical check that measurement variation is minimised. Preferably, however, a 'Gauge R&R' study will be conducted. This measures the repeatability (how often the same operator achieves the same reading) and reproducibility (how close different operators come to achieving the same readings as each other).

2.1.3 Analyse
Having measured a suitable sample, analysis is carried out. In its simplest form this may start with a Pareto analysis to identify key inputs. A main effects plot will usually supplement this and possibly an interaction plot, as well as more advanced inferential statistical techniques. These may include t-tests, means, median or ANOVA tests, chi-square or regression analysis depending on the format of the data. The results of this analysis should indicate which input has the greatest influence on the output that requires improvement. In some instances a technique known as 'Design Of Experiments' (DOE) may be employed to produce powerful results from a reduced number of experimental repetitions. This technique uses Fractional Factorial techniques derived from Taguchi methods. (2)

2.1.4 Control
Having introduced changes to the prime input factor, measurement of the process continues and the analysis is carried out on a sample taken subsequent to the changes. This 'after' reading is used to assess whether sufficient improvement has been achieved (set against the objective in the original problem statement). If sufficient improvement is not realised the process loops back to the analysis stage and the key inputs are re-assessed. When sufficient improvement has been achieved the process is monitored for a period and any procedural changes required are implemented, before the project is closed.

2.2 DFSS
The following steps are taken in a DFSS project.

2.2.1 Identify
At this stage the requirement for the project is justified in terms of strategic and customer requirements. The scope of the project is also defined.

2.2.2 Define
The key customer needs are defined and the critical design requirements necessary to meet customer needs are determined. The link between the need and the requirement should be documented and preliminary target values set where possible. Key tools at this stage are Quality Function Deployment (QFD) and decision analysis methods.

2.2.3 Design
Several design concepts are generated and these are assessed against the criteria formulated in the 'Define' stage. Following thorough analysis the best concept is selected. The QFD analysis may be extended to help with this process or other decision analysis processes may be used such as Pugh Concept selection. The decision criteria and weightings used should be documented along with any assumptions that have been made. Potential risks should be

identified at this stage and a plan developed to mitigate these. Capability analysis should be employed at this stage, if possible, and DOE may also be used if appropriate.

2.2.4 Optimise

Having chosen the best design concept, the opportunities for defects should be minimised. All risks that have been identified in the Design stage must be addressed. Several tools may be applicable at this stage including Finite Element Analysis (FEA), Computational Fluid Dynamics (CFD), Design For Manufacture and Assembly (DFMA), Failure Mode Effect Analysis (FMEA) and DOE. The key to design optimisation is in understanding which of the design features are critical parameters and then to understand the inherent variability of these features. This will usually require the use of inferential statistical techniques. By setting a parameter target value we understand that we are setting a target mean. Then, when setting tolerances, this is done with an understanding of the process capability and with reference to Six Sigma quality standards. This will be covered in greater detail in Section 4 below, 'Achieving Robust Designs through the us of DFSS'.

2.2.5 Validate

The design is tested to evaluate how well it meets the design requirements identified at the 'Define' stage. Any assumptions should also be validated at this stage. Test results should be analysed with reference to any Capability and Reliability requirements that have been stated. The FMEA and risk analysis should also be updated at this stage. If applicable the need for Statistical Process Control (SPC) should be identified and this should be included as Manufacturing procedures are updated.

2.2.6 Evaluate

The project should be reviewed and documentation should be completed. This should include an assessment of how well the project has delivered in terms of the requirements defined at the 'Define' stage.

3   SUITABILITY IN A DESIGN ENVIRONMENT: CONVENTIONAL SIX SIGMA AND DFSS

3.1   Conventional Six Sigma

In most instances new products are designed with reference to a large number of overall requirements. These requirements are often conflicting and it would be counter-productive to improve a single measure with no consideration to other requirements. If we take the example of component 'weight' and mean 'life', it may be possible to make significant improvement to one at the expense of the other. As both considerations may be equally important to the customer it would be inappropriate to consider only one requirement. The Six Sigma problem statement is therefore not well suited to design projects.

In cases of product improvement it may be possible to specify a problem in terms of a single measurement that needs to be improved. If other design requirements are not considered, however, this may still be inappropriate and could lead to an uncompetitive design.

If we follow this through the rest of the DMAIC process we can see that measurements prompted by the problem definition may be inappropriate. Analysis and improvement will be carried out based on such data and will only be appropriate if a sensible parameter has been included in the Problem Statement. The control phase will also evaluate the project based on the parameter chosen in the problem statement.

Although some projects in the Design Office may benefit from the traditional Six Sigma methodology care must be taken to ensure that all design requirements are considered, as appropriate. This tends to limit the use of these techniques to process improvement rather than product improvement.

3.2   DFSS

In contrast to the 'Problem Statement' used in the 'Define' stage of a conventional six sigma projects, DFSS specifies that having established the scope of the project in the 'Identify' phase, all the design requirements should be identified during the 'Define' stage. The use of QFD at this stage should allow prioritised design requirements to be established from a list of customer requirements. The next step, 'Design' encourages that several design options be developed and assessed against the design requirements and for risks to be identified. One of the designs is selected and the justification for the selection is recorded. During the 'Optimise' stage, risks identified are mitigated and the capabilities of the processes involved are assessed as required. The 'Validate' stage provides confidence that the design requirements have been met and the 'Evaluate' stage ensures that documentation has been completed as appropriate. These steps fit neatly with Design Office procedures.

4   ACHIEVING ROBUST DESIGNS THROUGH THE USE OF DFSS

As mentioned previously, the key to achieving robust designs is to understand the inherent variation encountered in parts used. Our design engineers are encouraged to measure actual parts to determine this variation in a statistical sense. In the past we often looked at the mean and probably the maximum and minimum values in a set of data. Furthermore, in some cases we would look to accommodate the worst possible stack of tolerances in terms of extremes. Our mantra has now become 'mean and standard deviation'. In most cases a normal curve can be determined from the mean and standard deviation and this allows sensible and achievable tolerance limits to be set. In cases where the required tolerances are too tight in relation to the observed spread of data, this allows the design engineer to consider alternative processes for part production early in the design, thus reducing cost. In terms of what is deemed an achievable tolerance, this will depend on the measurement, the part and its function. We encourage design engineers to look at making parts 'Six Sigma' by design (hence DFSS) where possible. The statistical measure of six sigma refers to tolerance limits set 6 standard deviations (σ) from the mean. Theoretically, this accounts for a 1.5σ drift which has been observed over time even in stable processes and returns around 3 parts per million as defective. These figures are based on research originally conducted by Motorola in the US.

See APPENDIX A (Origin of Parts Per Million defects against Sigma Levels)

Initially our design engineers stated that they could not measure some parts because they were new and had not yet been designed. The response in these cases was to ask if the new part was similar to a part that it replaced? If not, could they think of a similar part produced in a similar way by the probable supplier of the new part? Usually it is possible to find and measure a suitably comparable part without leaving the plant. In the rare cases where a suitable part cannot be found we have to resort to published figures.

The concept of 'Robust Design' comes from the work of Taguchi. The concept is involved with producing designs that naturally conform to the specified design requirements. Within the manufacturing environment that Taguchi was operating in, he was often faced with many different variables that could be controlled and his task was to find the optimum settings to produce a consistent product. To this end he developed his experimental techniques. For our product we assemble rather than manufacture. Our aim is still to produce a consistent product but it is usually more appropriate for us to develop our understanding from sample measurements rather than through experimentation. Further down the line we also use the Taguchi philosophy regarding 'loss function'. Taguchi developed the idea that all deviations from the target (mean) impact on the quality of the product. To this end we do not generally prioritise our Engineering Problem Notifications (EPNs) as we consider that it is often the small niggling problems that are neglected and difficult to fix. EPNs can be raised by anyone in the organisation and all must be closed satisfactorily. (3)

When building trucks we are commonly concerned with problems of 'tolerance build-up'. We have often used a worst-case approach, looking at the maximum and minimum values from a sample or from the tolerances set. This assumes that dimensions within the tolerance range are equally likely and can result in assemblies that are a 'looser fit' than desired. This approach may be desirable where there are few parts (less than four as a guide) or where the worst case may have safety critical implications. We now encourage our designers to use statistical tolerancing. Although there is a degree of risk in this approach, this is mitigated by understanding the inherent variability of each component as previously described. It is also stressed that this method is not appropriate if the theoretically possible stack-up creates a safety critical outcome.

By developing an understanding of the inherent variation that we are designing against, we believe that development lead-times will be reduced. By the use of simulations we can predict the outcome of expected 'tolerance build-ups' and determine at an early stage if the consequences are acceptable. This allows changes to take place while the design is still at the concept stage and exists only in the virtual world of the CAD model. It is obviously much cheaper to make changes at this stage in the design process.

See APPENDIX B (A Demonstration of Tolerance 'Stack-Up' Simulation).

Also, in cases where problems are encountered in production, the application of DFSS techniques helps us to resolve these problems. Having investigated the problem and after establishing the inherent variability of the processes involved we advise our Design Engineers that they have 3 options to resolve the problem. These options are; live with the problem, change the specification or improve the process. The choice to live with the problem can sometimes be justified when the cost of improving the process will cost more than the savings realised. The option to change the specification is usually applicable in cases where DFSS has not been fully utilised in the design process initially. On these occasions the original reasons for the specification being set at a particular limit are questioned. If there are good reasons for the specification to be controlled to the specified limits then it is inappropriate to change them. If, however, the specifications have been set in an arbitrary fashion, or they are tighter than they need to be, then it is possible to relax the specification to achieve an increase in capability (i.e. sigma level capability). Once again, the priority is that the Design Engineer understands the variability in the process, the relevance of the specification limits and the interaction between the two. The final option is to consider why there is unacceptable variation in the process and to either improve or replace the process.

See APPENDIX C (A Demonstration of Coping with Variation)

Through the use of tolerances that have been developed with an understanding of the inherent variability the probability of part conformance is increased. This, in turn, means that the probability of parts in an assembly fitting together without problems increases and consequently the proportion of products built 'Right First Time' increases.

5   SUMMARY

Although conventional Six Sigma techniques are good for problem solving and process improvements they are not well suited to the Product Creation Process in place in a design environment. Design For Six Sigma (DFSS) techniques are ideal for the design environment and promote the creation of 'Robust Designs' by developing an understanding of the inherent variability encountered in components and processes utilised. This understanding is developed through the use of statistical techniques. The creation of Robust Designs leads to a reduction in development lead times and a greater probability that the final product will be built 'Right First Time'.

REFERENCES

1:   Leyland Trucks Limited, Quality Procedures:
     QP4.2 PRODUCT CREATION PROCESS (FOR MAJOR NEW MODEL PROJECTS) –
     Issue 4, 20th January 2002

2:   System of Experimental Design – Genichi Taguchi (American Supplier Institute, 1977)

3:   Leyland Trucks Limited, Product Development Quality Procedures:
     EPO19 ENGINEERING PROBLEM CONTROL – Issue 15, 17th August 2000

Thanks to Tom Gargiulo, Nigel Thomson, Chris Griffiths and Steve Rothwell.

APPENDIX A
ORIGIN OF PARTS PER MILLION DEFECTS AGAINST SIGMA LEVEL

The following are examples of anomalies observed in some Six Sigma and Design for Six Sigma (DFSS) teaching material.

1) The explanation for Six Sigma begins with the following definition for 3 Sigma Level quality level:

NORMAL DISTRIBUTION

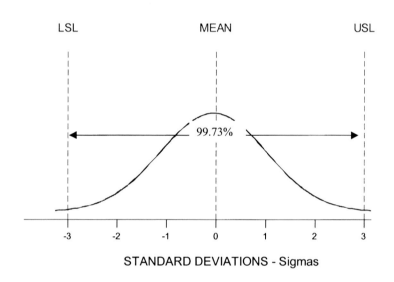

Figure 2.

Figure 2 implies that 0.27% of parts would fall outside the Specification Limits and therefore be classed as defective.

2) From the above, 0.27% defective parts will equate to:

0.0027 * 1,000,000 = 2,700 Parts Per Million (PPM) defects.

3) In the DFSS training material it follows, therefore, that for 3 sigma quality level the PPM defects are presented as 2,700.

In Six Sigma training material, however, 3 sigma quality level is presented as being equivalent to 66,807 PPM defects.

To explain how this figure is arrived at, we have to look to the next stage of the explanation of Six Sigma as a measure of quality. We also have to make several assumptions.

4) From 1) above the explanation continues that for many years (especially in the western world) this level of escaping defects (0.27%) was acceptable.

5) Following research based on many years of experience from real life processes, it was established that most 'good', 'well controlled' processes will tend to drift over long periods of time by around 1.5 standard deviations from the nominal position. This could be caused by many factors; examples might be tool wear, changes in personnel, changes in shift or movement of plant.

6) In reality this drift will differ from process to process, but from experience a figure of 1.5 standard deviations has come to be excepted as a good approximation for most 'good' processes (See Figure 3).

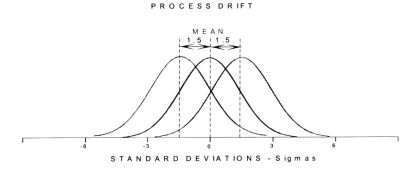

PROCESS DRIFT

Figure 3.

7) Assume that an initial sample was taken when the process spread was distributed about the nominal mean. Then assume that the process drifts 1.5 standard deviations from that nominal mean.

8) For a Six Sigma level design, this will equate to a process where the position of the mean =

$6 - 1.5 = 4.5$ standard deviations from the Specification Limit

From this (using z-tables used in DFSS training) we can derive the 3.4 PPM defects quoted in Six Sigma training material.

9) For a 3 Sigma level design, this will equate to a process where the position of the mean =

$3 - 1.5 = 1.5$ standard deviations from the Specification Limit

From this (again using z-tables) we can predict that the likelihood of parts falling outside the Specification Limit =

6.6807%

0.066807 * 1,000,000 = <u>66,807 PPM defects</u>

10) This is the figure for PPM defects quoted in Six Sigma training material for a Sigma Level of 3.

11) It will be noted that this is includes escaping defects falling outside one Specification Limit only (i.e. Upper or Lower). It is assumed that the process can only be losing parts outside one Specification Limit during the short term. This follows from an assumption that to drift from one extreme to the other (i.e. from +1.5 standard deviations to –1.5 standard deviations) will only happen in the long term.

12) The analysis is not 100% robust. It could be argued that over the period of 1,000,000 parts the process may see long term drift. The analysis follows the assumption that for most processes this shift + and – will not be seen. In reality it will probably depend on the individual process.

13) It will be stressed in DFSS training that each time a sample is taken the value of the mean and standard deviation for the population are predicted, but will always be wrong. The only way to obtain the true mean and standard deviation would be to measure all items in the population. As this is usually not practical we have to live with the information that we derive from a sample or multiple samples. This is still much better than guessing without taking any measurements.

14) The very simple explanation for the sample size of 30 being a good 'rule of thumb' is that this gives a level of statistical confidence that will usually be acceptable. Ultimately the level of statistical confidence derived from a sample size of 30 depends on the characteristics of the sample. 30 is only a 'rule of thumb' if it is possible to measure more, then the level of statistical confidence will increase, conversely if it is not possible to measure 30 then the level of statistical confidence will decrease.

     C606/012/2002 © IMechE 2002

APPENDIX B
A DEMONSTRATION OF TOLERANCE STACK-UP SIMULATION

We encountered problems with the clearance between the cab and radiator on a Pilot Build vehicle when tested for durability using an extreme surface on a test-track.

We started by measuring the gap between the radiator and cab-tunnel for the vehicle on test and for several vehicles on the production line. We established that 24 out of a sample of 29 vehicles had a gap less than the nominal minus the tolerance. Having established that the gap was too small we needed to establish the cause and how much variability was inherent.

The following were possible reasons to be investigated:
- Radiator wider than specified
- Cab tunnel narrower than specified
- Position of cab and radiator incorrect relative to each other
- Stack of extremes of tolerance

Using a portable Co-ordinate Measuring Machine (CMM) we undertook a further measurement exercise. We selected 30 vehicles (a variety of left and right hand drive) and took 14 measurements on each vehicle.

To allow us to assess the position of cab and radiator we took all measurements from a datum on the vehicle chassis. We soon established the following
(see Figure 4, where MU = Mean and SD = Standard Deviation):
- Radiator: close to nominal with little variation (i.e. small standard deviation)
- Cab tunnel: slightly narrower than nominal (Nominal 666.6mm, Measured Mean = 664, Measured Standard Deviation = 0.72)
- Radiator to frame: close to nominal with little variation
- Frame to Steering Column Hole (Cab Datum), less than nominal (cab over to right), significant variation
- Steering Column Hole to Tunnel Wall, greater than nominal (wall over to right), little variation

From this we could determine that the position of the cab and the position of the tunnel in the cab tended to add together to reduce the gap.

Finally we needed to predict statistically what the gap would be based on these means and standard deviations. To do this we needed to produce a 'Loop Diagram', relating everything back to the frame datum (See Figure 5).

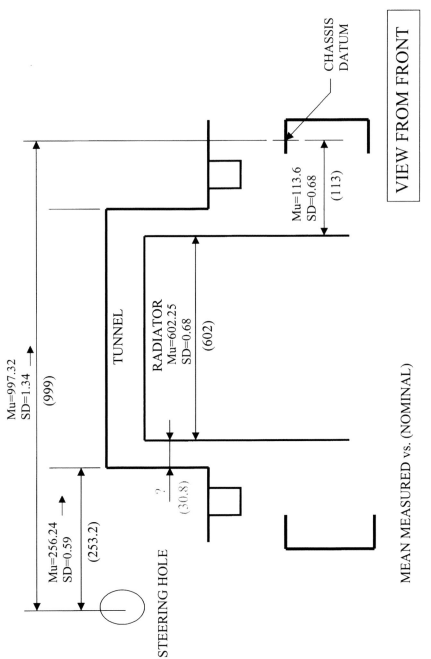

VIEW FROM FRONT

CHASSIS DATUM

Mu=113.6
SD=0.68
(113)

Mu=997.32
SD=1.34
(999)

TUNNEL

RADIATOR
Mu=602.25
SD=0.68
(602)

Mu=256.24
SD=0.59
(253.2)

?
(30.8)

STEERING HOLE

MEAN MEASURED vs. (NOMINAL)

Figure 4. STATIC CLEARANCES (SAMPLE n=18, Right Hand Drive)

C606/012/2002 © IMechE 2002

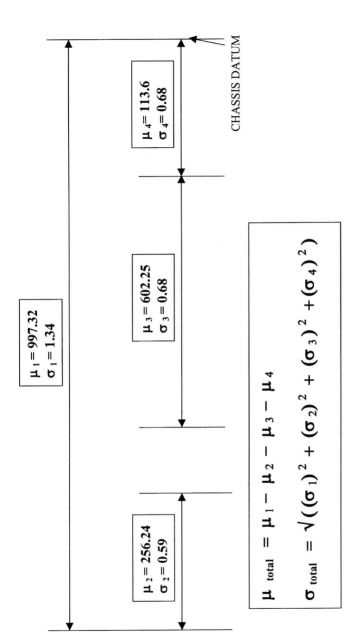

Figure 5. LOOP DIAGRAM

$$\mu_{total} = \mu_1 - \mu_2 - \mu_3 - \mu_4$$

$$\sigma_{total} = \sqrt{\left((\sigma_1)^2 + (\sigma_2)^2 + (\sigma_3)^2 + (\sigma_4)^2\right)}$$

From the formulae shown we can predict that gap will be 25.23 +/- 1.75 mm.
This compares with a nominal gap of 30.8 +/- 2 mm.

Whereas the gap is approximately 5mm less than nominal the variation is predicted to be less than the tolerance set.

To provide a visual representation, random data was generated and used to predict the process capability (see Figure 6).

Capability Analysis for Cab Tunnel to Radiator Gap (simulated)

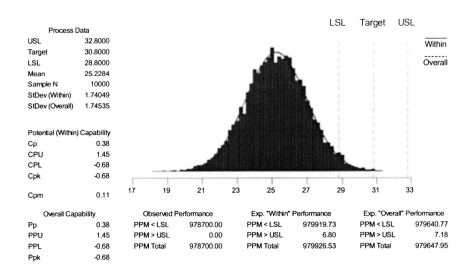

Figure 6.

Figure 6 shows a simulation of the tolerance stack that could be expected for this build using current processes.

The result of this investigation was to re-design the radiator mounts to cope with a static clearance of 20 mm to achieve a 3 sigma design.

It is interesting to note that before this investigation the perceived problem was that the radiator was out of position and that the solution would be to shift the static position of the radiator.

APPENDIX C
COPING WITH VARIATION

The position of a hole in a major component seemed incorrect. We were unsure whether to make changes to our parts to accommodate the position. Before we made changes we needed to understand whether the suppliers process was stable and in control. To this end we undertook a measurement exercise using a CMM and concluded that the parts were outside tolerance but that the process was in control. We presented our findings to the supplier who undertook a similar measurement exercise. Their findings concurred with ours.

We found that one important dimension was 2.2mm forward of the nominal. With a standard deviation of 0.99. This equated to the following capability with regard to the tolerances specified (Figure 7):

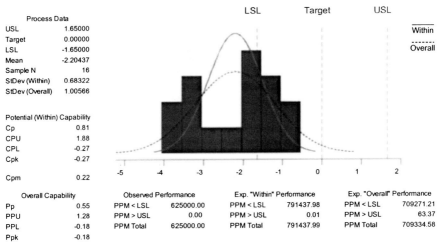

Figure 7.

We thus determined that less than 30% of holes would be within tolerance and that virtually all would be forward of the nominal. The process was running at 0 sigma level. If the whole process were to be shifted between the tolerances while maintaining this spread, or these tolerance were shifted to fit the process it would be running at 2.43 sigma level (deemed acceptable in this case).

The measurements from which the above results were derived were taken over a period of 22 days. Those readings were in turn taken some five months after our readings. As both sets of results seem to be generally consistent we had a reasonable degree of confidence that the process for controlling the position of the holes was stable and in control.

Having supplemented these statistical findings with a visit by our Supplier Quality Assurance (SQA) department, we agreed that the supplier could change the nominal position shown on their drawings. We then amended our parts to fit the actual position of the hole.

Index

Statistics for Engine Optimization

Edited by S P Edwards, D M Grove, and H P Wynn

In a series of specially commissioned articles, this book demonstrates how statistically designed experiments can make a major contribution to meeting existing and future demands in engine development.

Engine development teams are facing increasing demands from industry, legislators, and customers for lower cost engines and reduced vehicle exhaust emissions combined with improvements in specific vehicle performance and refinement. These necessitate the design, analysis, and testing of an ever wider range of options, while decreasing the process costs and/or duration.

Topics covered include:
Design of experiments; Modelling techniques; Response surface methods; Multi-stage models; Emulating computer models; Bayesian methods; Optimization; Genetic algorithms; On-line optimization; Robust engineering design.

Contents
● Editors' introduction: statistics for engine optimization ● The role of statistics in the engine development process ● Issues arising from statistical engine testing ● Practical implementation of design of experiments in engine development ● Statistical modelling of engine systems ● Applying design of experiments to the optimization of heavy-duty diesel engine operating parameters ● Engine mapping: a two-stage regression approach based on spark sweeps ● The application of an automatic calibration optimization tool to direct injection diesels ● An investigation of the utilization of genetic programming techniques for response curve modelling ● Using neural networks in the characterization and manipulation of engine data ● Empirical modelling of diesel engine performance for robust engineering design ● Improved engine design and development processes ● Subject Index.

1 86058 201 X   234x156mm
Hardback   208 pages   2000   **£49.00**

Professional Engineering Publishing

Orders and enquiries to:
Sales and Marketing Department
Professional Engineering Publishing Limited
Northgate Avenue, Bury St Edmunds, Suffolk IP32 6BW, UK
Tel: +44 (0) 1284 724384   **Fax:** +44 (0) 1284 718692
Email: orders@pepublishing.com
Website: www.pepublishing.com

Add 10% for delivery outside the UK

VISA   MasterCard

Advances in Vehicle Design

By John Fenton

This book is an introduction to vehicle and body systems. It provides readers with an insight into analytical methods given in a wide variety of published sources such as technical journals, conference papers, and proceedings of engineering institutions. A comprehensive list of references is provided.

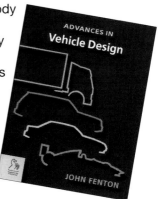

John Fenton distils and presents the best of this research and industry practice into an easily digestible, highly illustrated, and accessible form. Drawing on the available information, the author provides a well-structured and vital reference source for all automotive engineers.

Key features:

● Heavily illustrated ● Hundreds of examples ● Comprehensively indexed ● Concise synthesis of available information ● Written by an established authority ● A ready reference source

Contents include:

Materials and construction advances; structure and safety; powertrain/chassis systems; electrical and electronic systems; vehicle development; systems development: powertrain/chassis; system developments: body structure/systems; references; index.

1 86058 181 1   234x156mm

Hardback   190 pages   1999   **£64.00**

Professional Engineering Publishing

Orders and enquiries to:
Sales and Marketing Department
Professional Engineering Publishing Limited
Northgate Avenue, Bury St Edmunds, Suffolk IP32 6BW, UK
Tel: +44 (0) 1284 724384   **Fax:** +44 (0) 1284 718692
Email: orders@pepublishing.com
Website: www.pepublishing.com

Add 10% for delivery outside the UK

Integrated Powertrains and their Control

Edited by N D Vaughan
University of Bath

Integrated Powertrains and their Control
presents the latest advances in powertrain technology.
Specific applications are discussed, along with ideas as to how
concept, design, modelling, and simulation are put into practice, to
meet the latest/current legal requirements. The book will prove a
valuable source of information to all engineers and researchers working
on powertrains, fuel, control technologies, or emissions, and related
fields. Engineers will find this a useful volume as will vehicle
manufacturers and their suppliers and consultants.

The editor, Dr N D Vaughan is a senior lecturer in the Department of
Mechanical Engineering at the University of Bath. His principal interests
are in the application of control techniques to mechanical systems, the
design, of electro-hydraulic control valves, and the study of
continuously variable vehicle transmissions.

Contents include: Introduction to advances in powertrain technology;
Control of an integrated IVT powertrain; Driveability control of the ZI®
powertrain; Performance of integrated engine – CVT control,
considering powertrain loss and CVT response lag.

1 86058 334 2   234x156mm
Hardcover   110 pages   April 2001   £49.00

**Professional
Engineering
Publishing**

Orders and enquiries to:
Sales and Marketing Department
Professional Engineering Publishing Limited
Northgate Avenue, Bury St Edmunds, Suffolk IP32 6BW, UK
Tel: +44 (0) 1284 724384   **Fax:** +44 (0) 1284 718692
Email: orders@pepublishing.com
Website: www.pepublishing.com

Overseas
customers
please add 10%
for delivery